NEW WORLDS, NEW ANIMALS

Published in Association
with the National Zoological Park,
Smithsonian Institution

New Worlds, New Animals

From Menagerie to Zoological Park
in the Nineteenth Century

Edited by R. J. Hoage and William A. Deiss

WITH A FOREWORD BY MICHAEL H. ROBINSON

The Johns Hopkins University Press
Baltimore and London

To all Smithsonian employees, past and present, on the occasion of the Smithsonian Institution's 150th anniversary (1846–1996), for their devoted efforts to further the goals of the Institution.

This book has been brought to publication with the generous assistance of the Friends of the National Zoo.

© 1996 The Smithsonian Institution
All rights reserved. Published 1996
Printed in the United States of America on acid-free paper

05 04 03 02 01 00 99 98 97 96   5 4 3 2 1

The Johns Hopkins University Press
2715 North Charles Street
Baltimore, Maryland 21218-4319
The Johns Hopkins Press Ltd., London

Frontispiece: A Berlin monkey house, ca. 1884. For details see caption on p. 67.

Library of Congress Cataloging-in-Publication Data will be found at the end of this book.

A catalog record for this book is available from the British Library.

ISBN 0-8018-5110-6   ISBN 0-8018-5373-7 (pbk.)

CONTENTS

APPENDIXES

FOREWORD

Social institutions of all kinds evolve. This is true of political, legal, and administrative entities, as well as cultural, educational, and scientific bodies. The history of universities provides one such example. In the Middle Ages, they were principally centers for training men from the privileged classes for church and state leadership. They concentrated on classical languages and theology. Developing from this elitist base, they have become centers of broad-scale and diverse scholarship.

The history of zoos provides a similar example of historical transformation. Ever since our emergence from apelike ancestors, we have interacted with the living world in complex ways. We've spent the overwhelming majority of our existence in a precivilized state. After a very long period of hunting and gathering, we sapient primates started making collections of living things and altering them for our benefit. This was the origin of civilization, which was made possible by the domestication of plants and animals. The resulting division of labor and the accumulation of wealth produced tombs, temples, and collections of animals and plants. The period of civilization accounts for perhaps 1 percent of our history as hominids. With civilization came urbanization. Shortly after we had developed cities on a grand scale, zoos and botanical gardens sprang up in countries as far apart as Egypt and China.

It is difficult to determine which came first, the zoo or the botanical garden. Certainly there are records of royal collections of plants and animals dating from almost five thousand years before the present. The first zoo may have been a collection of several thousand animals in Saqqara, Egypt, around 2500 B.C.; the first botanical garden was probably the Shen Ming garden in China at about the same time. Both were prestige collections and probably only somewhat utilitarian. It did not take long, however, for the functional value of these two facilities to emerge. By 1500 B.C., the pharaoh Thutmose III had created a collection of medicinal herbs in his garden and thus established a connection between botanical gardens and medicine which has persisted throughout history.

Zoos have had, for much of their history, far less exalted goals. They began as, and continued to be, places of spectacle and entertainment. Research, education, and conservation are functions which, in the last one hundred years or so, have been grafted onto the recreational rootstock of zoos. For a long time, the only major change in zoos was in their audiences. They were originally imperial and exclusive, part of the appurtenances of wealth and power which distinguished many of the kingdoms of the past. Perhaps they symbolized dominion over beast as well as man; certainly they were ostentatious symbols of luxury.

Just as monarchs had "captive" poets and artists, such as Mozart and Handel, they also had menageries. The very term *menagerie* is presumably derived from the French *ménage,* with its connotations of household—in this case, of course, the royal household. Be that as it may, zoos flourished from the times of the pharaohs and Chinese emperors, through Solomon's collection of "apes and peacocks," down past the Roman circuses, William the Conqueror, and Henry II, to the Hapsburgs and others.

Botanical gardens, on the other hand, throughout their history, have maintained a close connection with science, first as centers for the study of herbal medicine, then as increasingly important areas for the study and propagation of plants with agricultural and commercial potential. Science and inquiry were largely anathema during medieval times; however, light did penetrate the darkness in places. Universities blossomed in the twelfth and thirteenth centuries, and with them came some scientific advancements. The major botanical gardens of Europe sprang up close to those universities that were centers of medical learning; later, even more important gardens developed in the capitals of the far-flung colonial empires.

From the start, exotic animals were fascinating to the peoples concentrated in north temperate regions. It was principally after the European navigators commenced their great oceanic voyages that new fauna and flora became a rich source of bioexhibits. Once the Age of Exploration was under way, leading eventually to colonization and empire building, temperate-region voyagers encountered the biological wealth of the tropics worldwide. As a result, major public zoos sprang up all over Europe and in North America. This process started in capital cities, spread into provinces, and then secondarily spread to the empires controlled from Europe. Zoos in Paris, London, and Berlin preceded Calcutta, Melbourne, Singapore, and so on. People flocked to zoos to see the wonders of nature.

Aquariums developed somewhat later, in the 1880s, after the glass tank was perfected. The breeding of oriental fishes in Asia had been mastered centuries earlier. Until fish could be viewed from the side, as in glass aquariums, the emphasis was on breeding ornamentals that were conspicuously colored from above—hence, goldfish, koi, and golden orfe. When modern aquariums began to be established in the late nineteenth century, they were frequently on the edge of the sea, because of a fixation with marine wonders and proximity to sealife. Only later were they located inland.

During the Age of Exploration and European colonization of the world,

The parallel histories of zoos, aquariums, botanic gardens, and natural history museums. (Courtesy of *Defenders* magazine, November/December 1987)

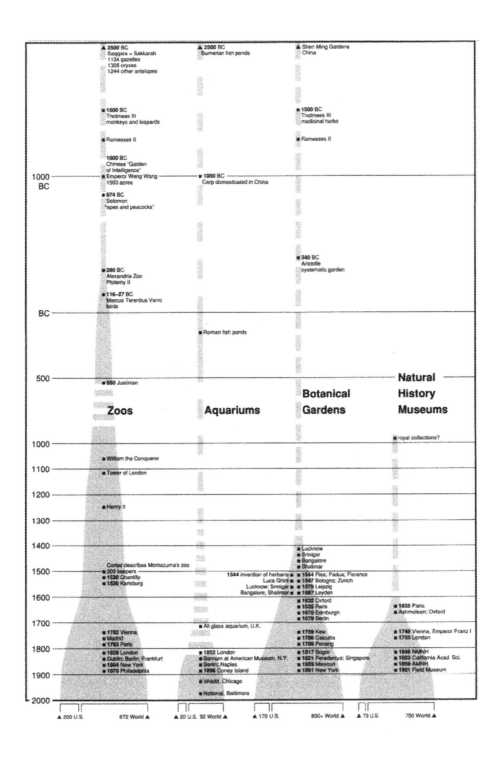

museums of natural history were established under the same impetus as zoos. They developed first around royal collections and later to store dead specimens of plants and animals for taxonomic purposes, while publicly displaying the marvels of life's past and the glories of biological structures. They stored and displayed nonliving specimens brought back to their largely northern homelands by explorers and explorer-naturalists. Bones, skins, pickled specimens, and dried plants could be transported and maintained for study and display more easily than living plants and animals. Fossil collecting on a grand scale opened up the secrets of the past. Such museums were centers of research, particularly in taxonomy. Museums of anthropology and antiquities blossomed during this same period. Conquests stimulated scholarship. For example, Napoleon plundered Egypt for its treasures; some of the booty still resides in Paris.

During the Industrial Revolution, education in the West was gradually extended to the majority of the population, and with it came a blossoming of curiosity; political, economic, philosophical, and scientific societies sprang up all over Europe in the latter half of the nineteenth century. The ventures of Darwin and others into exotic lands and the new flora and fauna they uncovered captivated the masses. Natural history books were printed in large editions. The evolution of zoos is inseparable from this process of cultural and scientific efflorescence.

As the scientific expansion of the nineteenth century proceeded, zoos became nominal centers for research, but in-house research was minimal. At that stage the natural world seemed inexhaustible, and its exploitation was blessed by religion and the prevailing philosophies of the age. The zoos of this expansionist period reflected the spirit of museum collectionism. They reflected the post-Linnaean passion for classification. Zoos labored to exhibit a comprehensive range of species; they also competed with one another to exhibit as many different kinds of animals as possible. This led not to natural groupings of individuals but to displays of solitary representatives of highly social forms caged in isolation.

When the National Zoo of the United States was established in 1889, its aims were declared to be "the advancement of science and the instruction and recreation of the people." The functions of the modern zoo are all but anticipated in that congressional mandate. The word *instruction,* with its authoritarian and didactic overtones, reflects an awareness that zoos could be educational. All that is lacking in this policy statement is reference to conservation.

At the beginning of the twentieth century, Carl Hagenbeck's revolutionary designs at his Tierpark, outside Hamburg, advanced animal exhibition beyond the bars, and later advances in animal health care and husbandry eliminated the sterile public bathroom style of indoor enclosures. Zoos developed habitat exhibits and finally passed from collectionism to natural social groupings of species and then ultimately to mixed-species exhibits.

The trend that Hagenbeck started evolved, at the end of the nineteenth century and beginning of the twentieth, into the BioPark concept, which the National Zoo has begun to implement (its recently completed Amazonia exhibit exemplifies this notion). A full-blown BioPark will combine elements of existing zoos, aquariums, natural history museums, botanical gardens, arboretums, and ethnological and anthropological museums to create a holistic form of bioexhibitry. Essentially, a bioexhibit should portray life in all its interconnectedness. Creating the BioPark means ending a series of unnatural separations. It means exhibiting plants and animals, not plants *or* animals. It also means that aquatic organisms belong in the same exhibition complex as terrestrial ones. Exhibits about the structure, functional interpretation, and history of life on earth belong inseparably with exhibits of living plants and animals. The BioPark concept creates a holistic perspective that corrects the accidents of our culture's history which caused natural history institutions to be separated.

The National Zoological Park will continue to transform what remains of its nineteenth-century-like "classical zoo" exhibits, based on taxonomic groupings, into replicas of functioning ecosystems (i.e., combining mammals, birds, fish, reptiles, invertebrates, myriad plants, and even the impact of humans) to form a holistic display. These exhibits will provide to zoo visitors a realistic view of how life forms and habitats are truly interrelated on this planet.

MICHAEL H. ROBINSON
Director, National Zoological Park,
Smithsonian Institution

PREFACE AND ACKNOWLEDGMENTS

New Worlds, New Animals: From Menagerie to Zoological Park in the Nineteenth Century is a unique book. Most of the chapters are derived from papers delivered at a symposium entitled "History and Evolution of the Modern Zoo," which took place in October 1989 to celebrate the centennial of the Smithsonian Institution's National Zoological Park. The symposium brought together most of the scholars currently engaged in research on the history of zoos, menageries, and animals in captivity. This volume is thus a compendium of the current scholarship in the field.

Other books provide brief overviews of the histories of menageries and zoos. None, however, takes an in-depth look at the nineteenth century, when zoos underwent several transitions—from elaborate royal menageries for the elite, to public facilities that expressed the power and might of a nation, to institutions dedicated to public education, wildlife conservation, and research. This period directly reflects social evolution in the West—from royal rule to parliamentary systems; from imperialistic, colonial empires attempting to "tame" newly discovered continents to today's more egalitarian, conservation-conscious world order.

The first public zoos in nineteenth-century Europe reflected the power of colonial rule, but gradually an appreciation of the study of natural history became widespread. By the late 1800s, as imperialistic attitudes and policies resulted in the extermination of animal and plant species, this appreciation had evolved into a concern for wildlife conservation. In the United States, this concern first became apparent in the efforts of several eastern zoos to save the American bison.

New Worlds, New Animals: From Menagerie to Zoological Park in the Nineteenth Century begins with overviews of the history of zoos in antiquity and what they symbolized in the societies in which they developed (Kohlstedt, Hoage et al., and Veltre). Most of the subsequent chapters describe the origin and development of selected public zoos in France (Osborne), England (Ritvo), Germany (Reichenbach and Strehlow), Australia (Gillbank), India (Mittra), and the United States (Kisling and Horowitz). Zoos in these countries had very different origins, including a former royal menagerie that became an extension of a natural history museum; an institution founded by a learned scientific society; a fish market that became an animal dealership and then an important zoo; a facility that was originally conceived as a haven for propagating disappearing species; and a way station designed to acclimatize Old World domestic stock to a new continent.

The book also examines the origin and evolution of zoos in the United

States, especially in the decades from the 1830s through the 1880s, when popular traveling animal shows and circuses with attached menageries (Flint and Kisling) declined and the first public zoos, such as the Central Park, the Philadelphia, and the National Zoos, were founded. In particular, this volume takes an in-depth look at the establishment of the National Zoological Park and its physical and philosophical transformation beginning in 1889 and continuing into the twentieth century (Horowitz and Ewing). The National Zoo is considered to be the first zoo created to preserve endangered species—a seemingly modern concept which, in fact, originated with the National Zoo's founders in 1887.

Perhaps the major contribution of this book is that it underscores the value of zoos as scientific and educational institutions that train students and scientists and stimulate public interest in natural history and wildlife. The written and photographic archives of zoos are especially important, as they permit scholars and interested members of the public to examine and understand human-animal relationships in a bygone era (Edwards). British zoo historian Clinton Keeling, speaking at the National Zoological Park in 1989, described the much maligned zoos of a century or more ago as "unexpectedly successful places in which a wide range of species bred freely and often led long lives. . . . [In fact] many world longevity records for zoo animals were achieved during the last three decades of the nineteenth century." Between 1860 and 1880 the London Zoological Garden successfully bred spotted hyena, pygmy hog, hutia, giraffe, chevrotain, African buffalo, bluebird, sun bittern, giant skink, several species of fruit bats and phalangers, and many other creatures, observed Keeling, who attributed this success to the fact that the animals were "looked after, watched over, petted and played with in a way that has today almost disappeared." It is the diligent record keeping of zoological park keepers and curators in the past that provides us with this kind of insight. It is these kinds of archives on which many of the chapters of this volume are based. Such documents can be an abundant source of information, available with only a small effort to any student seeking a unique perspective on how humans perceived animals many generations ago.

This book will elicit from its readers new respect for and understanding of the role of zoos in history, especially in the nineteenth century. It is hoped that the reader will also gain an appreciation of the important challenges awaiting zoos in the future.

FINANCIAL SUPPORT FROM THE Friends of the National Zoo and the Smithsonian Institution made this symposium possible. The 1989 National Zoological Park symposium "History and Evolution of the Modern Zoo" was cosponsored by the Society for the History of Natural History, which is headquartered at the National History Museum, South Kensington, London.

We are grateful to Michael H. Robinson and Gretchen Gayle Ellsworth, director and associate director, respectively, of the National Zoological Park, for their advice and support of this symposium. The staff of the National Zoo's

Office of Public Affairs, Sally French, Margie Gibson, Marc Bretzfelder, Susan Haser, Virginia Garber, and Michael Morgan, contributed considerable time and energy to make this project a success. Editors Amy Weissman, Gillian Lugbill, Margie Gibson, and Jane Mansour spent many hours converting early chapter drafts into final copy.

OVERVIEWS

REFLECTIONS
ON ZOO HISTORY

*T*his book explores the history of zoological gardens, primarily in Europe and the United States. It is a pioneering effort to sketch out a history of menageries and zoos up to the early 1900s, highlighting a number of important topics and suggesting an agenda for future research on the history of live animal keeping.

The emphasis here is on the nineteenth century, and that is not merely by chance. Historians have identified this era as the "century of science" because of the pervasive support for and major advances in science. Within the study of natural history, this was the "golden age" of museum development and, by extension, zoo development as well. Explanations for the increasing activity and the visibility of exotic animals, whether single specimens, traveling menageries, or semipermanent establishments, become more explicit as their organizers solicited publications and private support. At least three explanations for the sustained and increasingly deliberate collections of animals, particularly in the past three centuries, are offered in this book.

First (and most often articulated by zoo advocates) are scientific justifications. New discoveries and theoretical explanations in zoology, botany, and natural sciences created a demand for research resources. Nowhere is this more clear than in the imperial networks of France and England, as described in this volume by Michael Osborne, Linden Gillbank, and D. K. Mittra, and, to some extent, in the German zoos described by Harro Strehlow and Herman Reichenbach.

Scientific initiative and enthusiasm sometimes obscured the equally important economic and political activities that permitted, and even funded, the acquisition of living animals. Expanding European empires made exotic wildlife more accessible and stimulated the interest of naturalists, acclimatizers, and agricultural breeders. Transportation and communication networks throughout the empires were able to stock curio shops, create numerous amateur societies at home and abroad, encourage collections of minerals, plants, and animals, and establish botanical gardens, museums, aquariums, and zoos.[1] As Michael Osborne and Linden Gillbank demonstrate, organized societies and government agencies solicited animals from colonial officers throughout the French and British empires. There was great interest in the economic potential of importing exotics to different regions of the world. Although independent travelers and enterprising sea captains brought home specimens for public display, the exchange quickly became more systematic and commercial.

A third explanation for the growth in the collection of living animals is rooted in somewhat more elusive cultural conditions. Themes of human authority and control, evident from the scientific revolution in the seventeenth century, were expanded from the physical world into the animal world. Ritvo argues in this volume that while the early association of wealth and status with the rare and exotic furthered the interests of wealthy patrons in animal collections, the popular enthusiasm for animals had another psychological base: collecting animals gave people a sense of superiority and control over nature.[2] As society became more urbanized, curiosity about animals and national pride also served those creating zoological parks.

Chronology and Change

Taken together, the essays in this volume provide us with data and a chronology through which we can assess the emergence of the earliest modern, large-scale zoos and discuss why some approaches disappeared while others were sustained and improved. The authors demonstrate major shifts over time, but they rarely find comprehensive and dramatic transformations.

The menagerie, in the United States, was entrepreneurial. The chapter by Richard Flint demonstrates that menageries in the 1800s were a potent capitalistic mixture of promotional advertising and public demand. Communities and individuals welcomed the brief visit of exotic but potentially dangerous beasts to their towns. Small towns could be serviced by enterprising itinerant showmen, who needed to keep moving along wilderness roads in order to get sufficient return on their incredible investments (imported animals often cost nine hundred dollars each—a price that was double or triple the annual salary of a professional in that period). While some advertisements suggested an educational value in the exhibitions, that rhetoric seems to have been used more to suggest a distinction from circus entertainment than to indicate any explicit teaching function. In fact, animals on display could have been only of superficial and passing interest to the avid amateurs who took classification very seriously and sought to find norms rather than eccentricities in nature. Still, for several decades, the menageries seem to have attracted a large audience in the young United States.

The reason for the decline of menageries may help us place their history in context with that of the larger, more structurally complex, and often publicly supported zoos that came into being in the last half of the nineteenth century. The explanation may in part be due to changing tastes and the limited number of truly distinct species—what had once been exotic had now become familiar through display and popular publications. P. T. Barnum, whose meteoric rise to fame coincided with the decline of menageries in the 1840s and 1850s, was a master of innovation and humbug, constantly changing his standard exhibits at the American Museum in New York City as well as in his circus shows. While such shows drew popular audiences well into the twentieth century, they were challenged by natural history museums and illustrated popular magazines that began to attract an educated audience ready to learn more detail about animals in their natural habitats and their taxonomic relationships to one another.[3]

European education and cultural trends that popularized informal schooling and public leisure activity in the arts, music, and science quickly crossed the Atlantic. Harriet Ritvo and Michael Osborne describe how active zoological and acclimatization societies built elaborate facilities near London and Paris. Similarly, Herman Reichenbach and Harro Strehlow point out that the mid-1800s were a time for imaginative experiments in private and public patronage as well as in new kinds of animal collection and exhibition facilities on the European continent. Vernon Kisling Jr. and Helen Lefkowitz Horowitz suggest conscious emulation and exchange among those developing zoological parks and aquariums. If the pattern was similar to that in natural history museums, American historians will undoubtedly find that experienced European keepers and trainers arrived on our shores to staff these newer zoological parks.[4]

The best networks for exchanging ideas as well as fauna and flora were among naturalists. They shared a desire for the zoo to be a resource for scientific work. They distanced themselves from income-producing menageries, yet the histories here suggest that circumstance mitigated against any absolute or permanent break with that tradition. The power of private patronage and the centralized authority of France and England were considerably different from the diffuse American scene, where cities competed with one another and with the still-young capital in Washington.

Thus it was more national progressive sensibility among certain leading scientists at the Smithsonian than local civic pride that led to the establishment of the National Zoological Park.[5] Michael Lacey pointed out that leading intellectuals and scientists shared a common notion about governmental responsibility for environmental concerns which made a national zoo possible.[6] Facing issues of species extinction and conservation of unique or rare land forms, these leaders were committed to preserving certain individual animals and, at the same time, introducing a broad cross section of the populace to conservation education.

Indeed, late-nineteenth-century zoo administrators emphasized the relationship between environment and species, sought to investigate the behaviors exhibited by different species, and were part of a broader movement intended to enforce the humane treatment of animals. They began to recognize as problems the confined cages of menageries, the emphasis on individual specimens, and the perceived arbitrary (and sometimes even cruel) treatment of animals.

In the late 1800s, new trends brought larger cages that housed animals in groups or families—often with a fanciful eye to human habitat which might or might not evoke something of the relationships among animals in the wild. By the early twentieth century, a return to the parklike setting of a hundred years earlier, in which animals had a certain degree of freedom and space, signaled a new stage in zoo development, at once new and old—innovative yet with a clear tradition. The contributors to this book suggest, therefore, that there is ambiguity in terms such as *progress* and *improvement*. Their essays reveal that zookeepers have built on the experience of their predecessors. Continuity is also illustrated by the species of animals emphasized, the publicity and advertising that have been effective, and even the persistence of all forms of animal keeping, including the roadside menageries of reptiles and independent aviaries.

Historiographical Themes

The general history of zoos has not been well documented. An older, anecdotal kind of history which disparaged earlier activities in order to highlight innovation is challenged by zoo historians such as Vernon Kisling. We should remember that reasonable and humane treatment of animals has existed simultaneously with some unspeakable brutalities. We need to revise those sometimes self-serving insiders' histories of individual zoos written to enhance their own reputation at the expense of earlier endeavors.

Another important historiographical aspect this volume illuminates is the multiple and sometimes conflicting motives of those who established zoos. When zoos are most important, they are controversial and generate lively debates about the functions they might fill. Should they be primarily for "pure" scientific research, as a certain French director hoped? Should they deliver a message about conservation, as William Hornaday intended? Could they be used for applied research in domestication and agriculture, as acclimatizers anticipated? Were they primarily a moral amusement for a working-class population, or a source of education for the middle class? Research to date indicates that zoos were all of these things but rarely any one exclusively.

Depending on time, place, and financial resources, the popularity of these differing aspects of zoos waxed and waned. It is essential to take the discussion about zoo history beyond explicit debate by contemporaries and to investigate how zoos fit into the larger world. Zoos relied on and, in not so subtle ways, reinforced ideas of imperialism and authority, not only in nineteenth-century Europe but also to some extent in North America, where zoo collections featured animals from the American West, such as Rocky Mountain goats and Alaskan black bears, which conjured up visions of Manifest Destiny.

This volume gives us a significant source of fundamental information on the development of zoos, but it also raises new questions. It suggests some of the needs and opportunities for further research in zoo history. What happened in the twentieth century to zoos in the United States and abroad? Was there a hiatus with regard to new facilities and institutional innovation early in this century? What was the role of education, so much touted but so little described? What assumptions were held about the transfer of knowledge and what efforts were made to make the zoo/museum experience educational for children and adults? What international links were forged, beyond sales or exchange of animals, between the various museums? What was the role of national and international zoological associations? How did commercialism intersect with activities in popular public zoos? How did zoo development relate to the establishment of public transit lines? When were restaurants, book shops, and other commercial activities introduced? The recent centennial of the National Zoological Park has led to an examination of its history as well as certain historical aspects of other zoos, but clearly we still need more careful studies of particular institutions, more comparative studies, as well as comprehensive accounts of other aspects of animal exhibition.

There is a complex and changing relationship between visitors, patrons, and administrators that may not be readily discerned by looking at newspaper accounts and public statements of prominent museum directors. Moreover, zoos have, over time, tried to balance educational, scientific, and environmental concerns. Some of this history can only be recovered using visual records and artifacts—and thus it is essential that the photographs, canvases, and other physical records be identified and preserved alongside traditional printed institutional records.

Zoos provide a revealing perspective on human culture, a chance to observe the relative importance of other species and to integrate that information as we plan for the future of the planet. In effect, these chapters let us watch humans watching over animals during much of history. The accrual of information and skill, handed down in some cases through family lines and in other cases through apprenticeships and education, establishes a visible continuity from ancient times to contemporary zoos. At the same time, changing professional and public expectations challenge static or simplistic notions about the value and function of zoos. Thus, for example, the rather individual perception of William Hornaday about the disappearance of the bison a century ago has now exploded into a general public awareness of many species rapidly disappearing from the planet. Ecology, environmentalism, and biodiversity have become major themes in zoological parks as the twenty-first century approaches.

MENAGERIES AND ZOOS
TO 1900

Constructing a brief outline of the history of menageries and zoos has been highly revealing. There is no one comprehensive volume to which one can refer. At the beginning of the twentieth century, Gustave Loisel published much of what was known at the time about the origins of menageries and zoos. But considerable knowledge has been added to the field since the early 1900s. In *Animals for Show and Pleasure in Ancient Rome,* George Jennison supplied one of the more informative books about animal keeping and menageries in ancient Greece and Rome. A more recent volume on a closely related subject, J. M. C. Toynbee's *Animals in Roman Life and Art,* is equally informative. All these books are exceptional because of their detailed descriptions of animal collections and their purposes and because of their wonderful citations of original source material.[1]

Useful studies of zoo history also include Lord Zuckerman's *Great Zoos of the World,* Bob Mullan and Gary Marvin's *Zoo Culture,* and Stephen Bostock's *Zoos and Animal Rights* (chapter 2: "4,500 Years of Zoos and Animal Keeping"). Other recent books that are, at least in part, devoted to zoo history all suffer from a lack of properly referenced material—original sources of information are rarely listed.[2] F. E. Zeuner, *A History of Domesticated Animals,* also has some limited, but pertinent, material. Hernando Cortés's *Letters from Mexico,* Bernal Diaz del Castillo's *Discovery and Conquest of Mexico, 1517–1521,* and William Prescott's three-volume *History of the Conquest of Mexico* provide (or at least cite) original sources of information on the strange and spectacular imperial Aztec menagerie.[3] In the following outline, we have tried to cite original sources to the extent possible. Secondary sources that list primary materials are more common. In any case, every citation in this chapter is given a full bibliographic reference. This brief overview should give the reader a basic chronological framework in which to place the establishment of most of the zoos mentioned in this volume—the majority of which were founded before 1900.

Menageries from 2500 B.C. to A.D. 1800: The Realm of the Royal and Wealthy

One of the earliest wild animal menageries on record appears in Egypt. Beginning about 2500 B.C., pictographic and hieroglyphic records at the Saqqara cemetery near Memphis show that the Egyptians kept numerous cattle (as many as three species) and many oryx, gazelles, and other specimens of antelope. Other images at Saqqara depict Egyptians keeping such species as cranes, baboons, spotted nose addax, storks, brown addax, pigeons, ibis, and falcons. Egyptian rulers at the time apparently kept many wild animals in the attempt to tame or domesticate them. A number of animals were holy to the Egyptians; many were used in religious ceremonies. Baboons, falcons, and ibis took on special significance—specimens were found mummified and entombed at Saqqara. Carnivorous species kept included hyena, cheetah, and at least one species of mongoose.[4]

In the fifteenth century B.C., Queen Hatshepsut, daughter of Thutmose I of the Eighteenth Dynasty, sent history's first recorded wild animal–collecting expedition down through the Red Sea to the "Land of Punt" (presumably Somalia) to bring back monkeys, leopards, exotic birds, wild "cattle," and even a giraffe. These animals became part of the palace menagerie. Loisel labeled it the "first known Acclimatization Garden" (a place where the attempt was made to tame or domesticate wild animals for human use). Many exotic animals reached the Egyptian kings in the form of tribute. A wall painting from about 1330 B.C. shows antelope, cheetah, a giraffe, and monkeys being delivered to the pharaoh.[5]

Hundreds of miles to the east, lions were kept in pits or cages by the kings of Ur beginning about 2000 B.C., and bas-reliefs from the Indus River civilization of Mohenjo-Daro indicate that the Asian elephant was domesticated, harnessed, and put to work in the twenty-fifth century B.C.[6] Still farther east, in China, a nine-hundred-acre walled "Park of Knowledge" was constructed sometime around 1150 B.C. for Emperor Wen Wang in the province of Henan, which lies between Beijing and Nanjing. In the park were at least two kinds of deer, various birds, and an enormous quantity of fish.[7] Subsequent emperors possessed similar parks.

The Old Testament of the Bible indicates that King Solomon was interested in exotic animals and probably had some kind of palace menagerie in the midtenth century B.C. Passages from the first book of Kings state that Solomon's "throne had six steps, and [on] the top of the throne . . . on either side . . . two lions stood. . . . And twelve lions stood there on the one side and on the other. . . . [T]here was not the like . . . in any kingdom. . . . [In addition] the king had at sea a navy. . . . [O]nce in three years came the navy . . . bringing gold, and silver, ivory, and apes and peacocks." Passages from the second book of Chronicles repeat the above statements almost verbatim. Solomon must have known something about the keeping and breeding of large numbers of animals. The first book of Kings also states that he had "forty thousand stalls of horses for his chariots" and considerable amounts of "barley also and straw for the horses and dromedaries."[8]

According to Loisel, the ancient Greeks never had large animal collections comparable to the royal collections elsewhere in the ancient world;[9] however, the Greeks developed a more serious, inquiring attitude toward animals. From the seventh century onward, wealthy Greeks began to import and keep monkeys. In the following century, they experimented with captive francolins (quail relatives), acclimatized cranes, and purple gallinules; they also introduced domestic cats from Africa.[10]

In the fifth century B.C., the Greeks kept domesticated Egyptian geese and African guinea fowl, pigeons, chickens, ducks, and the remarkable Indian peafowl, which proved to be so popular that the public came from miles around to view them. They even paid for the privilege—the earliest recorded instance of a gate fee to enter a menagerie or zoo.[11] This animal collection was still in existence in the fourth century B.C. when young intellectuals from nearby schools were taken there as part of their general education.[12]

Traveling animal shows with dancing bears and lions and other animals were popular in ancient Greece. The Greeks kept and displayed wild animals associated with religious cults near temples in various city-states; historical records indicate that the exotic animals were sometimes tamed sufficiently to be used in religious processions.[13] As noted above, the Greeks kept animals for educational purposes and not just as religious or ceremonial symbols. In the fourth century B.C., Greek animal collections enabled Aristotle to write the first systematic zoological survey, *The History of Animals,* which describes about three hundred vertebrates known at the time. His student, Alexander the Great, sent information and specimens back to Greece from his eastern campaign. Alexander's travels east into Persia opened up a new source of animals for Grecian and other early menageries.[14]

Ptolemy I of Egypt (323–285 B.C.), who had a particular interest in natural history, founded a great zoo in Alexandria. His successor, Ptolemy II (285–246 B.C.), enlarged the zoo and made it the natural world's equivalent of the city's library. He also staged one of the largest animal processions in historical records. On the Feast of Dionysus in the year 285 B.C., the parade included 96 elephants drawing chariots, 24 lions, 14 leopards, 14 oryx, and, it seems, 16 cheetahs and 14 wild asses, among hundreds of domesticated animals. The procession lasted for most of a day.[15] Loisel writes that this procession did occur, but probably during the reign of Ptolemy VI (181–145 B.C.), and Howard Scullard asserts that it was Ptolemy II who first established the zoo at Alexandria and supplied it with elephants.[16]

In the Middle East, the last important king of Assyria, Ashurbanipal (669–

626 B.C.), inherited a zoological collection and was known as an expert on camels and lions.[17] In addition, falcon breeding and falconry were well established in Assyria by Ashurbanipal's time.[18] Nebuchadnezzar, king of Babylonia (605–562 B.C.), was a collector of lions, and, like the Egyptians, the Babylonians kept and trained large predators that they hunted for amusement.[19] This form of entertainment could have been a precursor to the bloody spectacles of the Roman era. Persian royalty, up to at least the fourth century B.C., kept large hunting reserves and animal parks. In the Babylonian region of the Persian empire, these reserves were known as *paradeisos*; they contained many kinds of animals including lions and panthers. The precursors of such reserves may have been the extensive but confined animal parks containing freely roaming lions and herds of gazelles maintained by Assyrian kings as early as Ashurnasirpal II (885–859 B.C.).[20]

By the second century B.C., wealthy and influential Romans were keeping private aviaries, fish ponds, and menageries to show their power and prestige.[21] Monkeys were found in these private collections as early as the end of the third century B.C.;[22] later snakes, gazelles, antelope, and other animals were kept.

Wild animals, especially leopards and lions, began to be imported for gladiatorial games as early as 186 B.C.[23] African imports, largely all for the arena, included ostriches, oryx, hippos, and crocodiles. Northern European and Asian animals included bears, bison, elk, and tigers. Lions, leopards, and elephants were brought in from both Asia and Africa. Most of these beasts used in bloodsport were received as gifts from governors of Roman colonies, foreign dignitaries, and monarchs; however, many others were obtained through animal traders.[24]

In Rome there were extensive animal holding areas, called vivaria, associated with the arenas that citizens might view. These were menageries of a sort. In the third century A.D., one, owned by the state, was located just outside the Praenestine Gate.[25] This vivarium was some 70 yards wide and 440 yards long, with one of its walls adjoining the city wall.[26] The enclosure seems to have been nothing more than a large rectangle; there was little attempt to make it a showplace, although the animals apparently could be viewed by the public.

Each Roman emperor had a menagerie for triumphal processions and official celebrations, especially gladiatorial exhibitions in which animals were killed. Octavius Augustus, Roman emperor at the dawn of the Christian era (29 B.C.–A.D. 14), had over 3,500 wild and tamed animals from his collection killed in twenty-six celebrations, including 420 tigers, 260 lions, 36 crocodiles, a number of elephants and rhinoceroses, and a snake reported to be 25 yards long. Emperor Trajan (A.D. 98–117) had 11,000 wild and tame animals slain.[27] Loisel examined the records of the menageries of Roman emperors between Augustus and Trajan: Caligula (A.D. 37–41) kept 400 bears and 400 African beasts, including camels; Claudius (41–54), 300 bears, 300 African beasts, 4 tigers, and many bulls; Nero (54–68), 400 bears, 300 lions, and many elephants; Titus (79–81),

5,000 wild beasts; and Domitian (81–96), rhinoceroses, elephants, lions, tigers, and bears.[28]

In A.D. 325, the emperor Constantine declared the arena games illegal, but Emperor Justinian (527–65) issued a decree again legalizing them.[29] By the end of the sixth century, arenas for blood sport appeared all over Europe.[30] With the decline of the Roman Empire, the arena events also declined, except in Constantinople, where the bloody games continued up to the twelfth century.[31] Remnants of these contests appear today in Europe and the New World in the form of bull fights, cock fights, badger baiting, and dog fights.[32]

In the thirteenth century, Marco Polo described the royal menagerie of Kublai Khan in Shang-tu. There was a huge marble and stone palace adjacent to a wall that surrounded many miles of parkland where game animals of all sorts were kept for hunting. Along with the game animals, Kublai maintained an animal collection that included leopards, lions (they were probably tigers, since they were described as having black stripes), lynxes, and even elephants, as well as a variety of eagles and falcons used for hunting.[33]

In 1417, Yung-lo, a Chinese Ming dynasty emperor, organized an expedition to Africa simply to obtain a giraffe. Apparently the emperor had a menagerie that included a zebra and other African animals.[34]

When Hernando Cortés and his soldiers reached Mexico City (Tenochtitlan) in 1519, they discovered that the Aztec emperor, Montezuma, kept magnificent gardens, a great aviary, and one or more structures that housed wild animals.[35] These facilities required hundreds of gardeners and animal keepers; there were three hundred attendants for the aviary alone. The aviary seems to have been vast, with dozens of species of birds, including many kinds of large and small birds of prey, parrots, pheasants, and perching birds. Numerous ponds, some of salt water, were devoted to ducks and other waterfowl. Cortés reported that the raptors, "from kestrel to eagle," were kept in a separate house equipped with indoor and outdoor cages. The animal collection apparently contained a large number of carnivores, including great cats, wolves, foxes, and smaller predators. Many of them were said to have bred. Diaz del Castillo wrote that in the animal collection were vipers and poisonous snakes which had on their tails "things that sound like bells." "These are the worst vipers of all, and they keep them in jars and great pottery vessels . . . and there they lay their eggs and rear their young."[36]

Diaz del Castillo suggested that the mammal and reptile predators were given as food not only "deer and fowls, dogs and other things" they hunted. "I have heard it said that they feed them on the bodies of the Indians who have been sacrificed."[37] The enormous aviary was occupied by

> birds of splendid plumage . . . assembled from all parts of the empire . . . the scarlet
> cardinal . . . golden pheasant, the endless parrot-tribe with their rainbow hues, . . .
> and that miniature miracle of nature, the humming-bird. . . . Three hundred atten-

dants had charge of this aviary . . . and in the moulting season were careful to collect the beautiful plumage, which, with its many-colored tints, furnished the materials for the Aztec painter. . . . A separate building was reserved for the fierce birds of prey; the voracious vulture-tribes and eagles of enormous size. . . . No less than five hundred turkeys . . . were allowed for the daily consumption of these tyrants. . . . [There were also] ten large tanks, well stocked with fish, [that] afforded a retreat on their margins to various tribes of water-fowl, whose habits were so carefully consulted, that some of these ponds were of salt water. . . . [In addition to the aviary, waterfowl ponds, and menagerie, there was] a strange collection of human monsters, dwarfs, and other unfortunate persons. . . . Such hideous anomalies were regarded by the Aztecs as a suitable appendage of state.[38]

This imperial menagerie ceased to exist after the Spanish conquest.

During the Middle Ages in Europe, the urgent need for large numbers of wild animals declined (the bloody Roman spectacles had diminished). Exotic animal collections became closely associated with royalty and the rising class of wealthy merchants. In addition, animals were used again as diplomatic gestures in the form of tributes, bribes, reparations, or ceremonial gifts.[39]

Charlemagne received exotic animals for his collection from other states, as did several later European monarchs, such as Holy Roman Emperor Frederick II, Henry I of England, and Philip of France. In 1235, Frederick II established at his court in southern Italy the "first great menagerie" in western Europe. At one time or another Frederick II had an elephant, white bear, giraffe, leopard, hyenas, lions, cheetahs, camels, and monkeys; however, he was especially interested in birds and studied them sufficiently well to write a number of authoritative books on them. Frederick II even took camels, monkeys, and leopards with him to Worms for his marriage ceremony.[40] From Henry I's menagerie at Woodstock in Oxfordshire grew the collection of Henry III, which the latter transferred to the Tower of London in 1252.[41] Apparently, the king pressured the citizens of London to provide financial support for his collection, but some of the animals, such as the polar bear and elephant, could be viewed by the public, and people came from miles around to see them.[42]

Trade and exploration during the Renaissance brought Europe into contact with new and wondrous worlds: Africa, the Americas, and Asia. This resulted in a resurgence in the importation of wild animals to Europe. Improved trading connections supplied many new zoos and menageries with rare and exotic animals. Royalty, affluent citizens, and merchants kept exotic animals in their gardens; traveling showmen took menageries from town to town for public entertainment.[43] The Renaissance nobility was not especially interested in New World species; it was the animals from Africa and the Far East that were most sought after.[44]

The Renaissance brought a more refined lifestyle to Italy, France, and other European countries which included acclimatizing (today we might call it habituating or taming) exotic animals at noble residences. The new species were considered useful; they were introduced into royal parks and game reserves for adornment and hunting. Many exotics, considered delicacies, graced the dining tables of nobles. Examples of these useful species included pheasants, guinea fowl, peacocks, bison, storks, herons, parrots, and ibex.[45] At this time, deer parks and wild animal reserves, long a privilege of the nobility in ancient and feudal times, started to appear in the form of urban menageries in European municipalities under the sponsorship of merchants and other citizens, for example, in Frankfurt in the late 1300s and in Augsburg and The Hague in the late 1500s.[46]

In 1450, René, count of Anjou and Provence, king of the Two Sicilies, and the duke of Lorraine, had a large menagerie at his château at Angers, possibly the most extensive in Europe before the menagerie established by Louis XIV of France in the seventeenth century. René had a lion house and enclosures for small mammals, ungulates, and ostriches. Also present were cages for songbirds, a large aviary, and a garden with a pond for waterbirds. The menagerie also contained dromedaries and even a "Moor" (who probably served as the animals' caretaker). However, the structures were scattered about the grounds; it was not a unified presentation.[47]

Leopards and especially lions were popular with European monarchs and princes throughout the thirteenth, fourteenth, and fifteenth centuries. Rulers of large and small realms all seemed to have a lion collection of some kind at one time or another, and gifts of lions occurred regularly between them.[48] This may be why European royal coats of arms, even today, often depict big cats.

Pope Leo X (1513–21), one of the Medicis of Florence, had a sizable menagerie at the Vatican which his family helped to develop. The collection included tropical birds, lions, leopards, bears, monkeys, civets, and one Asian elephant,[49] but he was not the first pope to collect wild animals. In the fourteenth century at Avignon, Pope Benedict XII apparently had two ostriches that came from the "Kingdom of Robert" (Sicily),[50] and Hahn reports that a lion house was established for the popes who lived in Avignon between 1309 and 1377.[51] Another of the Medicis, Cardinal Hippolytus of Florence, had a collection of "exotic peoples" as well as exotic animals. He is reported to have had a troop of "barbarians" which included Moors, Tartars, Asian Indians, Turks, and Africans.[52]

The sixteenth century witnessed the birth of menageries in urban centers across Europe and North Africa, including Prague, Karlsburg, Saint-Germain, Siena, Constantinople, and Cairo. The Schönbrunn Zoo at the Hapsburg royal summer residence outside Vienna has its roots in this era. In 1569, Emperor Maximilian installed a deer garden at the "Katterburgh" for the enjoyment of the imperial entourage. This animal facility evolved into the oldest extant zoo, Schönbrunn. After a period of major redesign and reconstruction, inspired by the menagerie built at Versailles in the 1660s by France's Louis XIV,[53] Holy

Roman Emperor Francis I presented the "new" Schönbrunn animal collection as a gift to his wife Maria Theresa in 1752.[54] Animals were obtained via royal expeditions to the Americas and Africa, and by 1770 the collection even included elephants.[55] The public was encouraged to attend, and the menagerie influenced the music, fashion, hair styles, and social life of Vienna to a remarkable degree.[56] The public was admitted intermittently to Schönbrunn, but it was essentially the private collection of the imperial family.[57] The Schönbrunn Zoo, even today, retains much of its 1752 design, and it is recognized as the first "modern" zoo.[58]

The royal menagerie, founded in Sweden in 1561, persisted into the eighteenth century when Carl von Linné, now known as Linnaeus, created the modern scientific system for naming animals and plants. His exposure to the animals in the royal menagerie is believed to have helped him frame his thoughts.[59]

Louis XIV established a magnificent menagerie in 1665 at Versailles, France. The facility, labeled by Loisel the first true "zoological garden,"[60] was maintained primarily through gifts to the king and featured 222 species at its zenith. Louis XIV created the world's first zoological garden that combined animals and plants in its exhibits. Although scientific research was conducted at the zoo, especially in the area of comparative biology, the animal exhibits were not scientifically arranged to educate the public about taxonomic relationships between animals.[61]

Unfortunately, Louis XIV's successor, Louis XV, was not interested in the menagerie, and the collection went into decline. During the French Revolution, an uprising by the people led to the release and slaughter of part of the collection. In 1793 the remaining animals were transferred from Versailles to the botanical garden in Paris, the Jardin du Roi, which Louis XIII had established in 1626 for the study of plants and natural history. Renamed the Jardin des Plantes, the plant and animal collections had become, by 1794, a division of the Muséum National d'Histoire Naturelle in Paris.[62] This was the world's first national menagerie. As it was part of the natural history museum, a close connection between the museum staff and the animal collection staff developed, which advanced the study of zoology within the institution. This museum-zoo association continued in other cities as new zoos were founded (Berlin and Washington, for example).

The Nineteenth Century: The Florescence of Zoos around the World

Up to the beginning of the nineteenth century, little had changed in Western menageries; nearly all were still the province of the nobility and the wealthy. Animals were caged for human amusement and as symbols of status and power. Exhibits were not systematically organized. With few exceptions, animals were displayed throughout the menagerie or zoo for "the gratification of curiosity and the underlining of the magnificence and the power of their owners."[63]

The incorporation of the Versailles menagerie into the Paris natural history museum (which was open to the public), the founding of the London Zoo in 1828 for "scientific" purposes (open to society members and guests), and the publication of Darwin's *Origin of Species* in 1858 all reflected a rising tide of public interest in the understanding and ordering of the natural world.[64] Most zoos established in the nineteenth century organized their exhibits based on the scientific classification of animals. Exhibit areas were often dedicated almost exclusively to primates, reptiles, carnivores, birds, ungulates, and so on. Only in the twentieth century have zoos constructed exhibits that illustrate ecological principles. Such mixed-species displays feature mammals, birds, reptiles, amphibians, and plants in any number of combinations.

Now that we have arrived at the nineteenth century, we defer this discussion to the many chapters in this volume which examine the establishment of zoos in Europe, the United States, India, and Australia through the 1800s and up into the early 1900s. In closing, we would like to list the opening dates of a large selection of zoos, primarily in Europe and North America. It truly was a time of florescence for zoos.

The records for the establishment of zoos in Europe and North America show that many other zoos could be listed. Similar records for the establishment of zoos on other continents are scarce at best. This list is given to indicate the blossoming of the zoological park movement in the mid- and late nineteenth century.[65]

1793   Jardin des Plantes Zoological Gardens in Paris incorporated the surviving animals from the Versailles menagerie.

1828   London Zoological Gardens opened to the general public. The Zoological Society was formed in 1826 and the gardens opened to members and guests in 1828. In 1829, more land was acquired, additional gardens and buildings were constructed, and 189,913 visitors came.

1833   Dublin Zoological Gardens opened in Phoenix Park. The Zoological Society was founded in 1830, "to form a collection of living animals on the plan of the Zoological Society of London."

1836   Manchester Zoological Gardens opened on property owned by John Jennison. Jennison had opened a small zoological collection in 1828 at Stockport. In 1836 he apparently transferred his collection to the Belle Vue location and opened a larger zoo to the public.

1839   Amsterdam Royal Zoological Gardens became a public zoo. In 1839 the Natural History Society purchased the menagerie of C. van Ascen and its property, which included buildings. In 1852 the king visited the gardens and bestowed on the society the name Royal Zoological Society.

1844   Berlin Zoological Gardens opened on part of the royal hunting grounds of Prussian king Friedrich William IV. It was a modest facility. In 1846 there were fewer than a hundred species in the collection.

1857   Rotterdam Zoological Gardens officially opened. In 1855, two railroad employees leased land along the railroad tracks where they kept exotic animals. Together with a group of animal enthusiasts, they formed a zoological society aimed at establishing a zoological park and botanical garden.

1858   Frankfurt Zoological Gardens opened to the public. The founding of the zoo by the city owed much to the efforts of the philosopher (and animal lover) Schopenhauer, who wrote, "Men are the devils on earth and animals the tormented souls."

1860   The Jardin Zoologique d'Acclimatation opened in Paris. This zoo was planned in 1855 by the Natural History Museum's Jardin des Plantes animal collection staff as an extension of the museum's zoo and was inaugurated by Emperor Napoleon III.

1860   The Cologne Zoological Garden opened. Plans for the garden were first made, and Dr. Heinrich Bodinus, mentioned in Strehlow's chapter, was appointed director in 1859. He left in 1869 to take the directorship of the Berlin Zoological Gardens.

1861   The Dresden Zoological Gardens opened in an area of the royal garden known as Poet's Walk which the king presented to the Zoological Garden Company. The zoo was first planned in 1859.

1863   The Hamburg Zoological Garden opened, adding Germany's first aquarium in the following year. The garden was first planned in 1860 by a small group of wealthy investors which founded the zoological society.

1865   The Breslau (now Wrocław, Poland) Zoological Garden opened. The garden was planned and construction began in 1863.

1866   The Budapest Zoological Garden opened. The garden was first planned in 1861. In order to encourage interest in the new facility, the zoo undertook acclimatization activities, including foreign domestic animals and cultivated plants.

1871   The Stuttgart Zoological Garden opened. The garden owes its origin to Johannes Nill and his animal collection. The garden's first design was created in 1866.

1872   The Royal Melbourne Zoological Gardens became a reality. The Victoria government first approved the establishment of a zoological garden in 1857.

1873   The New York Central Park Zoo was officially recognized. Beginning in 1861, a menagerie in Central Park served as a dumping ground for unwanted pets and carnival animals. Although the menagerie was open for public viewing, it was not officially recognized as a city institution until 1873, when it was incorporated into the Parks Department.

1874   The Basel Zoological Garden opened on land leased from the public infirmary until 1891, when the land was acquired by the state. The garden was planned by the Basel Ornithological Society in 1873 and at its opening housed about one hundred mammals and four hundred birds, nearly all of European or Alpine origin.

1874   The Philadelphia Zoological Garden opened. In 1859 the Zoological Society of Philadelphia made its first plans for the zoo. The American Civil War intervened, and the garden did not open to the public until 1874.

1875   The Cincinnati Zoo opened. The Zoological Society of Cincinnati was founded in 1873, and plans were first drawn up for a zoo.

1876   The Calcutta Zoological Gardens opened. The viceroy of India awarded a tract of land to the Honorary Managing Committee to establish the gardens in 1875. Private animal collections were immediately donated, and the gardens opened later in the same year.

1882   Ueno Zoological Gardens opened in Tokyo. Located near most of the city's scientific and cultural institutions, the zoo was originally administered by the Department of Agriculture and Commerce and later transferred to the Imperial Household Agency.

1888   The Cleveland Metroparks Zoological Park had become an operating public facility. In 1882 a herd of deer and property were donated to the city to establish a zoo. By 1888 at least one building and several enclosures were in place.

1889   The Atlanta Zoological Park opened. A prominent Atlanta businessman donated an animal collection to the city, and the public was given access.

1891   The National Zoological Park in Washington, D.C., opened. On March 2, 1889, an act of Congress created the National Zoo as a bureau of the Smithsonian Institution.

1892   The St. Petersburg Zoological Garden in Russia was already open to the public with at least one large building for pachyderms. Built on a site lacking much natural beauty, the zoo included a number of small theaters and cafés featuring live music.

1895   The Baltimore Zoo was an operating public facility. In 1876 the Maryland state legislature authorized the Baltimore Park Commission to establish a zoological collection.

1896   The Düsseldorf Zoological Garden opened. The collection, which included lions, tigers, polar bears, and a large aviary, provided competition for the nearby Cologne Zoo.

1896   The Konigsberg Zoological Gardens (formerly in East Prussia, now Kaliningrad) opened. The gardens developed from the Konigsberg Arts and Crafts Exhibition of 1895 and were connected to the city by an electric tramway.

1898   The Pittsburgh Zoo opened. Plans for the zoo were first made in 1895.

1899   The New York (Bronx) Zoological Park opened. The New York Zoological Society was founded in 1895.

1899   The Pretoria Zoo officially opened when the live animal collection of the State Museum of the South African Republic was moved. The new site had been used as a boys' hostel since 1895, but the outbreak of the Boer War left it vacant, as most of the boys joined the armed forces.

1899   The Moscow Zoological Garden was a functioning public facility. At the time it had many exhibits established. By the turn of the century, the collection included rare Przewalski's horses that were imported from Mongolia by the Hagenbecks.

MENAGERIES, METAPHORS, AND MEANINGS

*T*he establishment and use of a menagerie may be considered a material expression of deep-seated cultural values. While much of the existing literature in this field has examined the expansion of zoological science which has been brought about by the menagerie, it is my intent here to suggest some of the opportunities for greater sociological understanding provided by a study of the menagerie. My inquiry is related not to what a particular culture knows about animals but rather to what a culture's use of the menagerie says about itself. A menagerie is a small part of a larger process: that of a culture expressing its ideas about humanity through its relations with other animals and its environment. A menagerie reflects a culture's ideas about political power and ultimately the place of animals and human beings in the universe.

The term *menagerie,* commonly thought to be an old French word for "farmyard," is actually derived from the French root *ménage,* which means to manage, or management, and the suffix *rie,* which is used to indicate a place, as in *boulangerie* (bakery). In the literal sense, therefore, a menagerie is a place for the management of animals, a word that implies not only containment but, in a sense, domination and control as well. Also fundamental to the idea of a menagerie is the novelty of the animals it contains. Novelty can be in the form of either exotic animals from faraway lands or bizarre and freakish mutations of indigenous species that were saved from the fate of beasts of burden or becoming part of the food supply.

Because of their primarily utilitarian approach to animals, farms, ranches, and stockyards cannot be considered menageries. Unlike animals raised for food, or pet animals (which are usually treated more like members of the family), animals in a menagerie have been singled out to be unique representatives of their species. As such, all menagerie animals, from the freakish specimens of a circus menagerie to a royal menagerie's heraldic lions and eagles, evoke an emotional response that is different from any other encounter humans have with other creatures. A menagerie, whatever its physical form, is primarily concerned with the symbolic role of animals within a culture.

There are several prerequisites to the development of the menagerie. The existence of menageries in preclassical cultures was tied to the stage of economic and social development of the culture. It is obvious that a culture with an economic system based solely on subsistence agriculture would be disinclined to support a menagerie. The keeping of a menagerie requires a considerable investment of space, labor, and food resources—all for animals which, unlike domesticated species, are usually never eaten. Only cultures with a sufficient surplus, a rudimentary form of barter or medium of exchange, and a reasonably sophisticated form of government would have the wherewithal to consider establishing a full-fledged menagerie. In view of this, the act of keeping wild animals in early cultures may be considered an indicator of an advanced level of social development and a sign of economic stability.

The appearance of early menageries was concurrent with the rise of urbanization, and, with few exceptions, menageries have remained an urban phenomenon. Several factors may have contributed to this, including the complex labor demands of menageries, which must have precluded their establishment until the population of artisans and other skilled laborers had reached a sufficient size. Jennison noted that even the smallest Roman "vivarium" must have required a platoon of stone masons, blacksmiths, and carpenters to maintain the facilities, as well as warehouses full of grain, flocks of sheep and goats to provide fresh milk and meat, and an army of keepers, trainers, and attendants to transport animals for exhibition or use in games.[1]

There must be a certain element of novelty in the act of viewing animals. Until the development of cities, encounters with wild animals, while perhaps fraught with danger and crucial to the survival of the community, could not be considered a novel event. Perhaps the rise of cities created a nostalgia for the wild environment, and in response the menagerie evolved as the "art form" of encountering the wild.

In his definitive work on the history of the menagerie, Gustave Loisel divided the development of the menagerie and the subsequent rise of the zoological garden into five periods.[2] They provide an interesting basis for discussion, for each period appears to coincide with certain significant intellectual, political, and economic changes in Western culture.

1. *The Prehistoric Period.* There is evidence that prior to the development of systematic agriculture in the early Neolithic era, nomadic peoples caught young wild animals that were not intended as food.[3] These animals were kept tethered at the edges of camps and were probably used as objects of play or were killed and parts of their bodies used as elements of costume or decoration. Once habituated to people, these animals may have also been useful as decoys in hunting. In any event, to a hunter-gatherer people, the constant presence of a game animal would not have gone unnoticed, and a collection of such creatures

would carry a particular element of status. Loisel speculated that this might have been an early form of royal menagerie, which after the development of permanent settlements and the rise of agriculture led to the paradeisos.

2. *Period of the Paradeisos.* The Persian word *paradeisos* refers to a large, walled park where a large number of beasts were kept for the exclusive contemplation and enjoyment of the monarch. Paradeisos provided animals for royal hunts and ceremonial processions as well as storage space for animals given as tribute from foreign peoples. The earliest record of such a park comes from China (established around 1150 B.C. by Emperor Wen Wang), but evidence of such parks can be found in the empires of Assyria and Babylonia and later in the Egyptian dynasties.[4] This concept of the royal "paradise" (a model for the mystical "Garden of Eden") was sustained in the West until the fall of the Roman Empire, but the form survived in China well into the nineteenth century.

3. *Period of the Menagerie.* The word *menagerie* is applied by Loisel to a set of cages used to confine exotic animals, usually grouped by general class or family (i.e., all the felines together, all the primates together, and so on). In contrast to the paradeisos, the menagerie of the Middle Ages was not an attempt to create a private heaven on earth but rather was little more than a collection of living trophies kept on the palace grounds, a reflection of the ruler's importance and the extent of his empire. (The connection between menagerie and empire cannot be understated.) Loisel points to the Aztec capital of Tenochtitlan (Mexico City) as having the grandest menagerie of this period, containing a collection that was unmatched in the Europe of his day.[5] This zoological and botanical treasure was recognized as being central to the spirit of the Aztec empire—a fact that Cortés understood only too clearly when, as the first act of his violent campaign to overthrow Montezuma in 1521, he slaughtered the animals and set fire to the aviaries, demoralizing the Aztec people.

4. *Period of the Classical Zoo.* The aftermath of the 1789 French Revolution inspired the establishment of a zoo as a public education and recreation institution. The nineteenth century witnessed an explosion of these new institutions: London (1828), Amsterdam (1839), Berlin (1844), Antwerp (1843), Philadelphia (1874), and many others. Whether financed by public funds or private zoological societies (as in London and New York), these parks, dedicated to a combination of scientific advancement and public amusement, became a source of great civic pride.

5. *Period of the Modern Zoological Park.* A little more than a century after the revolutionary French menagerie, a new period in zoological exhibition was inaugurated in Germany by the famous animal dealer Carl Hagenbeck (1844–1913). In 1907 Hagenbeck, in collaboration with Swiss sculptor and architect Urs Eggenschwyler, constructed the first park to display animals in large "bar-less" enclosures surrounded by deep, barely visible moats at Stellingen, outside Hamburg.[6] The design of the exhibits attempted to re-create the natural habitat of the animals, promoting more naturalistic behavior and allowing the public to view

them in a context similar to their native environments. Although this approach has spread slowly, it is considered the dominant mode of modern zoological exhibition throughout the world today.

While the physical manifestation of the menagerie has experienced gradual, evolutionary changes, the role that these institutions have played in society has changed dramatically—as would be the case with any cultural institution that has survived more than three millennia. The Persian paradeisos and the modern zoological park might both contain animals in natural settings without bars, but there is a significant difference between the former's role as the royal "Garden of Eden" and the latter's as a multipurpose facility with recreational, educational, and scientific functions.

The changes in the cultural role of the menagerie have paralleled what communication scholars recognize to be the major changes in the communication forms of Western culture. The periods in the evolution of the menagerie could be considered a reflection of the new symbolic and epistemological environment created by each communication change. This is clearly evident in how each culture viewed its menageries and in the metaphors it used to describe them. In a broad sense, we can say that oral and preliterate cultures viewed menagerie animals as living words, symbols, or totems; scribal (preprint) cultures viewed animals as icons and living allegories; and print cultures viewed animals as books, turning menageries into living libraries of nature.

Animal Symbolism and Prehistory

Throughout history, the emotional relationship between humans and animals has been motivated by anthropomorphic projections: the assigning of human qualities to animals. The urge to anthropomorphize seems almost universal among cultures. Several theorists link this urge with the emergence of symbolic language. John Dewey speculated that the first "stories" of early language-developing hominids would revolve around experiences with animals and the hunt, reasoning that these were emotion-laden events central to the survival of the group. Dewey considered the assignation of human qualities to one's prey to be one of the most basic symbolic transformations and one that could easily be made even in the most simple of early linguistic cultures.[7]

On a more sophisticated level, Suzanne Langer points to anthropomorphism as a "symbolic shorthand" in the development of early religion. "A god who symbolizes moral qualities does well to appear in animal form; for a human incarnation would be too confusing. Human personalities are complex, extremely varied, hard to define, hard to generalize; but animals run very true to type. [Their qualities] . . . are exemplified with perfect definitiveness and simplicity by every member of their species." In Langer's view, animals fulfill an important need in oral cultures for complex, yet unambiguous, symbol systems. Animals are a living alphabet, and with the rise of totemism (the belief in an animal or object as the original ancestor and current sacred emblem of a clan or people), the symbolic role of the animal is absolute.[8]

Durkheim observed that "images of the totem-creature are more sacred than the totem creature itself."[9] A preliterate or oral culture's incipient menagerie can be seen as the transformation of individual animals into symbolic representatives of their species. Removed from the context of their natural environment and the need to exhibit instinctual survival behavior, animals in such early menageries were not accurate representations of their counterparts in the wild. Rather, they became symbolic reminders of nature which the culture wished to honor.

Scribal Culture and the Allegorical Menagerie

If the symbolic value of menagerie animals was recognized in preliterate cultures, then cultures with scribes producing written documents appear to have recognized that the symbolic communicative power of menagerie animals could be extended and amplified through the exhibition and exchange of specimens.[10] The closed paradeisos, originally developed as a private royal repository for totemic animals, became in the nineteenth century the open menagerie, a public display of these symbols. This is similar to the manner of early writing, in which the monarch's royal statements, once private, are made public by the display of symbols on buildings and monuments—proof of the monarch's status and power in that culture.

The gift of exotic animals in honor of a foreign-power leader has a history far older than the Nixon-brokered U.S.-China giant panda exchange of 1972. During the reign of the Assyrian king Tiglath-Pileser I (ca. 1100 B.C.), the pharaoh of Egypt sent the king a gift of a large crocodile, a "donkey of the waters" (hippopotamus), and several strange "fishes of the sea."[11] While this is the first recorded instance of a tribute of live animals, it is quite probable that this practice

The vulture represented a number of female deities at various times in Egyptian history, including Isis, Hathor, and Mut. The baboon represented Hapy, one of the sons of Horus. During different periods, the serpent represented different gods, including the sun god, Re, and his daughter, Maat. (Illustration by Vichai Malikul)

dates from the very beginning of complex societies. Through historical records, however, we find these "tribute animals" became a standard medium of international exchange and their display an important symbol of empire.

> The meanest chieftain in the poorest country of the uncivilized world could offer in this kind of tribute as fine as that as the most powerful monarch—a gift made more valuable by the glamour that clung and still clings to the distant and the unknown. The very presence of strange creatures, beauteous birds or ferocious beasts, was a living proof of the monarch's might and influence, so that it should not surprise us to find that from the earliest times . . . a very fine zoological collection has marked the crest of power in every nation, and shrunk with its fall.[12]

The Roman Empire best exemplifies the imperial use of animals. Animal games (*ludi*) and spectacles of animal slaughter (*ventaiones*) were given by successive administrations under the republic, each in an effort to break the records of previous events. Under the early empire, however, there was a veritable explosion of animal display. "On the whole, it would seem that under the early Empire (when compared to the Republic) more striking appeals were made both to the capacity of the spectators for sympathizing with animals or for admiring exhibitions of skill, and also to their taste for the sight of agony."[13]

Such sights were considered fitting for a martial people. Not only did they reinforce the idea of blood lust as popular entertainment, but by exhibiting such spectacles as battles between bears from Germany and lions from North Africa, they stood as vivid reminders of the power and extent of the empire. The propaganda value of animals was also employed in the provinces. Several elephants were brought to Britain during the Claudian invasion of A.D. 43, not as battle animals, but to accompany the emperor in ceremonies designed to frighten and impress the natives.[14]

The rise of Christianity at the end of the Roman era introduced new conceptions of animal symbolism. Medieval people saw themselves as the descendants of Adam, who was made in the image of God, for whom God created the earth, and over whose animals he had dominion. Lynn White has observed that the heritage of the creation myth has instilled in Western culture a unique view of nature and animals. "Christianity, in absolute contrast to ancient paganism and Asia's religions . . . not only established a dualism of man and nature, but also insisted that it is God's will that man exploit nature for his proper ends." The early church considered the study of nature as a branch of theology, for nature "was conceived primarily as a symbolic system through which God speaks to men; the ant is a sermon to the sluggards, rising flames are the symbol of the soul's aspiration."[15]

Owing to the economic decline of the West, the number and scope of the early Christian era menageries were undoubtedly modest, and records of "great" collections are scanty at best. Detailed knowledge of exotic animals (such as it was) could be found only through the writings of antiquity, generally

Aristotle or Pliny the Elder.[16] It is perhaps this lack of firsthand knowledge of animals which aided the popularization of the allegorical view of them. Without the contradiction of truth through observation, the symbolic role of animals was free to blossom in complexity and richness.

The Christian's need to see animals as the embodiment of God's messages, combined with the scribal tradition of annotation and amendment, created a literature notable for its symbolically exploitative approach to animals, as well as fanciful flights of imagination. To the Roman, the lion was considered the "king of beasts" simply because it was the "beast of kings"—an emblem of wealth and power well recognized in public exhibitions. In contrast, the late-second-century Christian manuscript "Physiologus" described the lion as having three characteristics. First, it erases its footprints with its tail so that hunters cannot follow it, signifying that the incarnation of Our Lord is unknown, even to the powers of heaven. Second, it sleeps with its eyes open, signifying that when the body of Christ "slept" in the death of crucifixion, his divine part was awake at the right hand of the Father. And finally, lion cubs are born dead, but after three days their father breathes into their faces, bringing them to life, thus reminding us of the passion, death, and resurrection of Christ.[17]

In the light of such belief, the keeping of a medieval menagerie could be considered not only a rare event but also one fraught with pious significance. At the same time, the menagerie retained its traditional secular role as a symbol of power and empire. With this powerful mixture of sacred and secular symbolism, it comes as no surprise to discover the establishment of many royal menageries in northern Europe stocked with animals brought back by noblemen returning from the Crusades.

Frederick II von Hohenstaufen, Holy Roman Emperor and leader of the Fifth Crusade, traveled most of his life accompanied by his menagerie, which (in a description from November 1231) included elephants, camels, panthers, gerfalcons, lions, leopards, and bearded owls.[18] The royal menagerie of England was established when Frederick presented King Henry III with three leopards, which were housed in the Tower of London and immortalized on the royal coat of arms.[19] Philip VI of France set aside a corner of the Louvre in 1333 to house his "*bestes estranges.*"[20]

The Menagerie in the Print Era

The impact of the print-based communications revolution of the fifteenth century stimulated the application of modern science to many subjects. The advent of scientific inquiry imposed a new logical order and brought status to the role of the menagerie among the learned. A detailed discussion of the effects of the printed book on the science of western Europe is far beyond the scope of this chapter, but there are some clear observations to be made about changes in the menagerie brought about by the printed book. Metaphorically, it might be said that scribal culture's menagerie animals were treated like stained glass windows in a cathedral, iconic reminders of stories to be contemplated, while print culture's menagerie animals were treated like books, to be studied, classified, and compared in minute detail.

There were, of course, attempts at zoological classification which preceded print. Most notably, Aristotle had devised a scheme that separated animals into two classes: "blood-bearing" and "blood-less" animals.[21] Several manuscript and early print works arranged animals in alphabetical order by Latin name (for instance, listing the donkey under "A" for Asinus and not next to what we would consider its relative the horse, which appeared under "E" for Equis), an order that in some cases was preserved regardless of the language in which the text was originally written or published. Once print was fully established, however, the "feedback and correction loop" of scientific publishing[22] encouraged increasingly more detailed and logical systems of classification, culminating in the classic "Systema Naturae" of Carl von Linné.

Embedded in the concept of the Linnaean system of classification is an important redefinition of the process of acquiring knowledge. Whereas previous authors concerned themselves with the description of the form and function of animals, Linné considered internal anatomy rather than outward appearance as his basis for classification. "A great advance too . . . was the segregation of the animals combined under the class of *mammals*. Popular prejudice was long universal and is still largely against the idea involved. Sacred writ and classical poetry were against it. It seemed quite unnatural to separate aquatic whales from the fishes which they resembled so much in form and associate them with terrestrial, hairy quadrupeds."[23]

To the general public, Linné's intricate system of classification by order, genus, and species seemed to contradict common sense. This was, of course, the failure to recognize that outward appearances do not represent the total reality—an issue that had caused such problems for Galileo a century earlier. The idea that the metaphorical "book of nature" was open for all to read was again being challenged. Galileo's celebrated response was that although the "book of nature" was open to public inspection, it was not "given to every man to know and read . . . it is written in the language of mathematics."[24]

Although appealing to the print culture, the scientific metaphor of the menagerie as a laboratory of life did not eclipse the previous public perception of its symbolic roles as allegory, empire, and totem. Even though the operation of the Menagerie of Versailles, established by Louis XIV, was under the direction of the Academie des Sciences (and produced the first important work in comparative anatomy), it was still first and foremost a political testament to the power and majesty of the king. Because of its tradition as a symbol of royal opulence, popular fury against the menagerie in the early days of the French Revolution led to its partial destruction in October 1789.[25]

The revolutionary French "National Menagerie of the Museum of Natural History" (established with animals from the former royal collection) was cited by

Loisel as the first classical zoological park. But aside from the change of ownership, this park, and the many others that followed it in the nineteenth century, differed little in symbolic content from their royal predecessors. For example, the Zoological Society of London was founded in 1826 for the purpose of "the advancement of zoology and the introduction into England of new and curious animals."[26] Not surprisingly, the "new and curious" animals consisted of lions left over from the closing of the royal menagerie and "Sumatran animals" collected by an officer of the East India Company. The menagerie's association with royalty and empire had not appreciably changed.

What had changed, however, was access. Whether available for free (as in Paris) or by subscription (as in London), the conception of the nineteenth-century public menagerie was not unlike that of a public library or museum— access to information and education for all. But unlike these newer institutions, the menagerie still resonated with centuries of symbolic meaning, helping to evoke an unprecedented degree of emotional involvement on the part of the public. Thus, the development of the classical zoo revived and gave new meaning to the historical symbolic roles of the menagerie.

With the new spirit of openness, zoos became highly competitive, each striving to exhibit the largest number or the most exotic species. Popular attention through the telegraph and newspaper made some animals into national figures. (This was vividly demonstrated by the public outcry in England when it was announced that P. T. Barnum had purchased Jumbo, the London Zoo's famous elephant, and was planning to ship it to America.[27] Jumbo was the pride of England, and his loss was considered nothing less than scandalous.) The popular press also thrived on stories of the great "bring 'em back alive" zoo men of the late nineteenth century. Men such as Carl Hagenbeck (and later Frank Buck) not only supplied zoos with ever more exotic specimens, but they did everything they could to encourage the aura of the "great hunter," which, in a sense, was little different from the totem bearer of the Neolithic era.

Despite the anthropomorphic hoopla, zoo professionals continued to use the print medium in guiding both the design of zoos and the use to which they were put. With the development of long, erudite, museumlike graphics, zoo-going was considered more than ever to be a literacy-based experience. In effect, the zoo was an encyclopedia of life, illustrated with live animal exhibits. The zoo was described as a vast storehouse of life, a living panorama, a compendium of biological knowledge. A popular belief evolved which held that a great zoo should be as comprehensive as possible, containing (like a good postage stamp collection) one of everything from all over the world.

The print era menagerie wrestled with its dual symbolic roles of science and showmanship. This dualism has produced some striking contrasts and to this day is a cause for serious debate among zoo professionals. (What other academic and research facility, for instance, sells popcorn, cotton candy, and often animal rides?) The scientific, social, and moral issues in the life sciences are becoming more sophisticated. While zoos are confronting the questions of genetic diversity and the forces of the international economy on habitat destruction in the Third World, they risk being less and less understood by the public. When zoo curators exhibit highly endangered snow leopards in order to discuss their desperate losing battle with humans for dwindling resources in the high mountains of Tibet, they must always contend with anthropomorphic visitors who will bring their children to "see the big pussycats."

Menagerie, Metaphor, and the Electronic Culture

In reading historical accounts of various menageries and zoos, I have found that the symbolic roles of these institutions are often revealed in the metaphors used to describe them. The ancient Chinese built contemplative "thought gardens," the Persians named their zoos "paradise," and few monarchs were without a collection of animals in tribute to the scope and majesty of their realm. Each era has developed new metaphors and meanings for the menagerie, reflecting the religious, political, or intellectual bias of its culture. At the same time, these new meanings altered the balance between the older symbolic roles of the menagerie as totem, allegory, and empire. It is my thesis that the symbolic meaning of animals in a culture, and the metaphors used to describe them, are a product of each era. As our culture is transformed from a print to an electronic medium, how will this transformation be reflected in the menagerie? Can we speculate what new roles will arise for the menagerie in an electronic culture? I would like to suggest two areas that offer possibilities for further exploration.

When the "bar-less zoo" approach to animal exhibition is used to display a number of different plant and animal species from the same geographic area, it attempts to present animals in an accurate, naturalistic environment representative of a particular ecosystem. This metaphor of exhibiting "little worlds" (such as the Bronx Zoo's JungleWorld) might be seen as a reaction to the decontextualizing effects of the electronic environment. Rather than a parade of this animal . . . and now this animal . . . and now this . . . , each complete "little world" is filled with interrelated meanings and messages about the great circle of living things. At least, this is how the zoo professional would like to think of it, giving the zoo visitor the full picture of the variety and complexity of every ecosystem.

This may not be the case, however. In the image-rich environment of the electronic age, the naturalistic, multispecies exhibit may be the zoo's attempt to compete with the vividness of television, an admission that a "bird in a box" exhibit will no longer interest a public raised for decades on "Wild Kingdom" and "National Geographic" spectaculars. The issue of context may no longer be valid in a culture that believes that the entire planet can be enjoyed from the comfort of one's easy chair.

The electronic culture's penchant for vivid imagery may also help create a

---

new symbolic definition for the zoo. Zoo-going has traditionally been a visual experience, a fact that was often in conflict with the print-based approach of the nineteenth- and early-twentieth-century menagerie. (One goes to *see* a zoo, not *read* it.) With the decline of literacy, as some have suggested, and the advent of "secondary orality," the zoo may be experiencing a revival as iconography. The general director of the New York Zoological Society uses iconlike metaphors in referring to the Bronx Zoo's Elephant House as "a palace in praise of elephants and rhinos" and its JungleWorld as "a cathedral of the diversity of the tropical rainforest." Like the medieval stained glass window, even icons in this electronic era are intended not so much to make one think as they are to make one feel. It is this almost religious feeling of awe and respect for wildlife and natural systems which is the goal of modern zoological exhibits.

Could it be perhaps that in a future filled with vivid images and devoid of contact with the natural world, the zoo will indeed become a place of worship? A cathedral filled with animal icons to remind us of the love we once had for a natural world long since gone?

European and Colonial Zoos

ZOOS IN THE FAMILY
The Geoffroy Saint-Hilaire Clan
and the Three Zoos of Paris

*N*umerous dynastic scenarios permeate the history of zoos and botanical gardens. Nepotism was common enough in these institutions to merit more than our passing interest. In commercial
zookeeping and the natural history trade, passing on the business to family
members made good economic sense. The practice also provides historical clues
to the nature of the zookeeping art and a window on zoo direction in an age of
protoprofessional management.

A few examples demonstrate the extent of this practice, which existed in
those halcyon years before World War I.[1] At Great Britain's Kew Gardens, the
director's chair was occupied by two generations of Hookers and handed on to
their in-law William T. Thiselton-Dyer.[2] In Australia, several generations of the
Le Souef family directed zoos in that country. Hamburg's Hagenbeck family has
provided the historian with one of the best examples of a zookeeping dynasty in
the nineteenth century, along with the Geoffroy Saint-Hilaires of Paris. Scientific
credentials were necessary for the best posts, but patronage and experience
carried considerable weight. Kinship played an important gatekeeping function
in determining who would become a zoo director.

However widespread the practice, my concern here is with three men and
three institutions. The most celebrated zoologist of the triad, Etienne Geoffroy
Saint-Hilaire, founded the menagerie at the Paris Museum of Natural History
during the French Revolution. This was the city's first public zoo, and Etienne
directed it for more than four decades. In 1838 he was replaced in this function by
his son, Isidore, who directed the menagerie until his death in 1861. Isidore
would influence the history of all three Parisian zoos. He was a motivating force
behind the city's second public zoo, the Jardin Zoologique d'Acclimatation. He
also participated in planning a third zoo, in the Bois de Vincennes, which is now
the city's major facility for keeping and displaying exotic animals. Isidore's only
son, Albert, served as his informal zoological apprentice. But he did not follow
his father and grandfather into the rarified ranks of the museum professoriate and

thus missed the chance to become a mini-industry for historians of French zoology. Nonetheless, Albert exerted considerable influence over the evolution of the Jardin Zoologique d'Acclimatation, which he directed for nearly three decades. As a scientifically informed entrepreneur, he sought to balance the requirements of running a public attraction with the desire to realize his father's utilitarian goals for practical zoology and rural economy.[3]

In a brief romp through the history of these personages and institutions, I hope to show how particular visions of zoology influenced the genesis and development of these three great Paris institutions. Any program of zoological research is the result of both institutional and intellectual constraints, and I conceptualize the relationship between institutions and scientific ideas as both reflexive and symbiotic. For example, the Paris Museum of Natural History's mission of public instruction combined with the uncertain fiscal health of its menagerie exerted a preponderant influence on the museum's options for zoological research. Conversely, a theoretically inclined philosophical zoologist like Etienne, who held simultaneous posts at other Parisian institutions, was reluctant to give high priority to a task such as zookeeping, which would not further his career.

Etienne Geoffroy Saint-Hilaire and the Museum Menagerie

The menagerie at the Paris Museum of Natural History owes its existence to the sweeping reforms of the French Revolution and the widely felt need for the museum's predecessor institution, the Jardin du Roi, to be more responsive to the needs of agriculture.[4] Although the Jardin du Roi contained more than six thousand species of living plants, the keeping and study of live animals there had been a rare occurrence. Naturalists such as Buffon and Daubenton had secured the Jardin du Roi's place in the history of zoology, but the vast majority

Etienne Geoffroy Saint-Hilaire (1772–1844), Chair of Mammals and Birds at the Paris Museum of Natural History, also served as the director of the museum's menagerie for more than forty years. (Image by permission of the Bibliothèque Centrale du Muséum National d'Histoire Naturelle, Paris)

of the live animals they studied had been housed on Buffon's estate at Montbard. The Revolution—and, more specifically, the Convention Nationale—changed this in June 1793 with the approval of a constitution for the Museum of Natural History which endorsed the creation of a menagerie and the incorporation of the Jardin du Roi, newly renamed the Jardin des Plantes.[5]

The new Museum of Natural History included two newly created chairs of zoology. The Chair of Insects and Worms was given to Jean-Baptiste Lamarck, who had been a botanist at the Jardin du Roi under Buffon. The Chair of Mammals and Birds was to have gone to Bernard-Germain-Etienne de Lacépède, who had written the volumes on reptiles and fishes for Buffon's *Histoire naturelle* and was widely regarded as Buffon's proper successor. Being of noble birth, however, Lacépède was forced to flee Paris in 1793, and the Chair was subsequently awarded to Etienne Geoffroy Saint-Hilaire, then twenty-one years of age.[6]

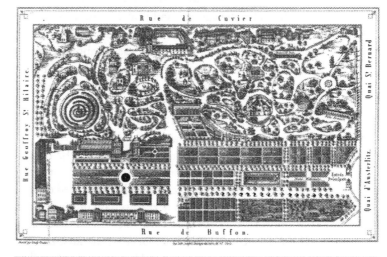

The Paris Museum of Natural History was founded by the Decree of the Revolutionary Convention of June 10, 1793, incorporating the Jardin des Plantes (formerly the Jardin du Roi) and endorsing the creation of a menagerie, all of which are detailed in this map from 1845. (Image from Pierre Boitard, *Le Jardin des Plantes* [Paris: J.-J. Dubochet, 1845])

A nineteenth-century postcard shows a panoramic view of the Jardin des Plantes with the city of Paris in the background. (Postcard courtesy of the Smithsonian Institution Library, National Zoological Park Branch)

The manner in which Etienne obtained animals for the menagerie provides an exemplary chapter in opportunistic zoo management. Although he knew occasional success in obtaining animals that he desired for the menagerie, a history of its growth reveals a collection built with little rational selection. In the fall of 1793, the police began to clear the streets of animal acts, spectacles, and other attractions that might incite a mob. They took the animals and their keepers to the museum, where the young professor, who acted without consultation from his superiors, accepted the animals and placed the keepers on salary as guardians.[7]

Confiscated animals continued to be added to the menagerie in 1794. In April they were joined by the handful of survivors from the royal menagerie at Versailles. Many of the Versailles animals had been released or butchered in the previous year. Among those that remained were a lion and his canine companion, a quagga (a now extinct zebra), and a rhinoceros, although the last died before it could be moved to the museum menagerie.[8]

France did not have much of an empire until after 1830, and any equation between the menagerie's cultural function and empire must be drawn with caution. Keith Thomas and more recently Harriet Ritvo have written on the significance of animal collections as symbols of empire.[9] Nonetheless, the menagerie's collections do provide a fair barometer of French diplomatic and commercial activities, as the collection grew through gifts from the dey of Algiers (1798), the emperor of Morocco (1824, 1832), and the Egyptian pasha Muhammad-Ali (1827). Additional animals were brought in after French army victories in Austria, Holland, and Algeria.[10]

Etienne directed the menagerie during the museum's so-called golden age, which ended around 1840. Most of the zoo's buildings were completed in this era; however, he did not seem overly concerned either with the kind of zoo which was built or with conducting research on the zoo's inhabitants. In the tradition of Buffon, Geoffroy Saint-Hilaire rejected the Linnaean system of taxonomy as arbitrary and artificial, and he became dissatisfied with the museum's mission to classify all of the natural world. He was unable to follow through with publishing a catalogue of living mammals and birds in the museum's collections.[11]

The menagerie's acquisition policy was only slightly more selective than that of Noah's ark. It collected pairs of animals, which were displayed, painted, drawn, classified, and perhaps ultimately dissected in the amphitheater of comparative anatomy. Acquisition of numerous kinds of animals pleased the menagerie's public, for they wanted the most bizarre and greatest variety of animals placed on display. But the policy was adverse to certain kinds of zoological research. The museum's menagerie was small, with only limited space; it did not have room for herds of animals. An intensive research program on the reproduction or improvement of quadrupeds, for example, was out of the question. Moreover, the elder Geoffroy Saint-Hilaire was not given to building a collection of animals which would support research on the improvement of French agriculture and industry, nor was he interested in the practical aspects of domestication, animal behavior, or experimental physiology.

Recent books on Etienne portray the scientist as moving further and further from empirical concerns around the time of Isidore's birth in 1805.[12] As the founder of the school of philosophical anatomy, his clashes with Georges Cuvier, his archrival, and brilliant flashes of theoretical excess placed him on the margins of French zoological science by the 1830s. Long before, in 1795, Geoffroy Saint-Hilaire had brought Cuvier to the museum, and the two had become friends and collaborators, sharing the common goal of elevating zoology beyond mere taxonomy to something more philosophical. Cuvier's adherence to Linnaean classification and his rapid success, which eclipsed Geoffroy Saint-Hilaire's own, made the two men rivals in an ongoing debate that ended only with Cuvier's death in 1832.[13]

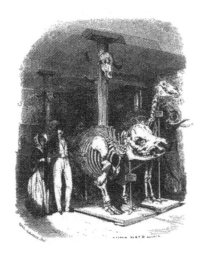

One might speculate whether this rhinoceros skeleton, on display in the Hall of Comparative Anatomy at the Museum of Natural History, belonged to the animal that died before it could be moved from Versailles to the new menagerie during the French Revolution. (Image from Pierre Boitard, *Le Jardin des Plantes* [Paris: J.-J. Dubochet, 1845])

These enclosures at the Jardin des Plantes depict two concurrent movements in zoo architecture. The ponderous elephant house, typical of early large mammal houses, reflects the need to display human dominance. The Moorish Algerian gazelle house, on the other hand, illustrates the use of architectural styles from the species' native lands. (Images from Pierre Boitard, *Le Jardin des Plantes* [Paris: J.-J. Dubochet, 1845])

Geoffroy Saint-Hilaire's colleagues at the museum were well aware of his neglect of the menagerie, and in 1837 they tried to remedy the situation by shifting responsibility for its control to a newly created Chair of Comparative Physiology for Frédéric Cuvier, the younger brother of Georges Cuvier. The arrangement lasted for less than a year and terminated with Frédéric's death in 1838, at which time Isidore Geoffroy Saint-Hilaire became interim director.[14]

Isidore Geoffroy Saint-Hilaire and the Three Zoos of Paris

Born at the family home at the museum in 1805, Isidore grew up to be a part of his father's professional and intellectual world. A zoologist and historian of zoos who later became director of the menagerie in 1841, Isidore had begun his career convinced that Bacon's book, *New Atlantis,* provided the blueprint for the ideal zoo.[15] The Baconian zoo was a multifunctional institution simultaneously serving to educate and amuse the public, promote the advancement of general zoology, and improve agriculture by the careful selection and study of animals. But utopian enthusiasms dwindled as the Second Empire aged. After two decades at the helm of the menagerie, Isidore jettisoned the multifunctional ideal and became France's foremost champion of specialized zoological institutions.[16] To be sure, this shift in Isidore's thinking paralleled the era's tendency to organize natural history into disciplinary units such as geology and ornithology,[17] but the institutional context should not be overlooked. The museum charter charged the menagerie to accomplish all the Baconian functions, but construction of new facilities had virtually ceased after completion of the new ape colony building in 1836. The 1850s and 1860s brought severe crowding for the animals, a political regime hostile to the museum, and zero budget growth.[18]

To Isidore, the solution lay in creating an annex of the menagerie which would assume some of its functions. In 1860, while serving as director of the museum, he proposed that a branch of the menagerie be built in eastern Paris in

Georges Cuvier (1769–1832), whose rivalry with Etienne Geoffroy Saint-Hilaire consumed the latter's attention throughout his tenure as director of the museum menagerie. (Image from Pierre Boitard, *Le Jardin des Plantes* [Paris: J.-J. Dubochet, 1845])

the Bois de Vincennes.[19] The menagerie was to remain a center for taxonomic endeavors and pure scientific research. The branch would provide what the museum had always lacked: a secluded environment where its animals could reproduce and be studied and raised under competent supervision. At the branch, wrote Isidore, animals would find conditions more favorable to "their fecundity, the rearing of their young, and the conservation of types in all their purity [and] even their improvement."[20] Although the museum obtained land for the facility, enthusiasm for the project waned after Isidore's death, especially when costs for the project were calculated at 1.1 million francs, about twice the then yearly budget for the entire museum.

In contrast to the planned Bois de Vincennes zoo, which would not be realized for decades, Isidore's efforts to create a specialized zoo on the western fringe of Paris met with rapid success. This zoo, the Jardin Zoologique d'Acclimatation, was developed in association with the Société d'Acclimatation. It became a reality when forty acres in the Bois de Boulogne were consigned to the Société by the city of Paris, along with a ten-acre gift from Napoleon III.[21] This zoo differed from the museum menagerie in many respects. It was not constrained by being a department of a natural history museum, as the menagerie was, and it owed little to taxonomic science or comparative anatomy. Furthermore, its large, state-of-the-art enclosures, modeled after the London Zoo, were designed to hold herds of a few selected animals deemed to be of real or potential economic value.

The founders strove to meet the needs of what Isidore Geoffroy Saint-Hilaire called applied natural history rather than pure scientific endeavor. In the words of its honorary president, Prince Jérôme Napoléon, the Jardin sprang from a desire to "exit the domain of theory to enter that of practice, and [to] place the results of our efforts before everyone."[22] When the Jardin opened in 1860, it was embraced by the Rothschilds, the emperor, and a host of the Second

The commercial atmosphere of the Jardin d'Acclimatation and its accent on entertainment are captured in this nineteenth-century postcard. (Postcard courtesy of the Smithsonian Institution Library, National Zoological Park Branch)

Empire's notables. The founders invested one million francs to open the Jardin, which was established to present only exhibits deemed useful to the public good. Their mission was to acclimatize, domesticate, breed, and sell new exotic animals that would meet the needs of industry, agriculture, and animal *amateurs*. Thus, Angora goats, llamas, alpacas, and their products were kept on permanent exhibition alongside active silk farms, weaving rooms, agricultural implements, and the like.[23] The institution went a good distance toward realizing the Geoffroy Saint-Hilaire family motto, "to be useful," and it became an arena for Isidore's transformist zoological theories and a test of the administrative skills of his son, Albert.

Albert Geoffroy Saint-Hilaire and the Jardin Zoologique d'Acclimatation

When Albert Geoffroy Saint-Hilaire completed his *bachelier ès sciences* in 1855, he was responding to his father's wish that he "seriously occupy himself with the sciences."[24] Albert had been involved in the planning and construction of the Jardin Zoologique d'Acclimatation, and he became its assistant director in 1860. He then moved up to the directorship in 1865. Albert, however, was more an entrepreneur than a scientist, and his term as director was marked by the institution's rebirth from the ashes of the Franco-Prussian War, its transformation from utilitarian zoo to profit-seeking attraction, and, finally, its slip into bankruptcy.

In the 1860s, the Jardin averaged about a quarter million visitors per year, held about five thousand animals (three times the number housed at the museum's menagerie), and usually managed to meet costs. But troubles surfaced the very year Albert became director, as the zoo experienced its first financial loss, a moderate shortfall of some fifteen thousand francs. The institution was closed during the Franco-Prussian War and the Commune of 1870–71, when it became

Albert Geoffroy Saint-Hilaire (1835–1919), as he looked in 1887 during his term as director of the Jardin d'Acclimatation. (Photo courtesy of Gérard Geoffroy Saint-Hilaire, author's collection)

home to 130,000 sheep and 20,000 cattle destined to feed a starving Paris. A number of the Jardin's animals became evacuees and rode the rails to Belgian zoos at Antwerp and Brussels. Those that remained, enough to fill thirty-five railway cars and sixty carriages, went to the museum menagerie, where most died, or were killed for food.[25]

Albert skillfully handled the evacuation and, to his credit, rebuilt the Jardin with financial assistance from the government and gifts of animals from other European zoos. The crowds returned, and in 1874 the Jardin entertained almost 600,000 visitors and cleared some 40,000 francs from the sale of animals worth about 300,000 francs. All was not well, however. The future of the zoo had been mortgaged by huge loans and shares on the Paris stock exchange. Worse yet, the display of practical zoological science was not very remunerative.[26]

In the 1860s, scientists, landowners, and the Bonapartes might be encountered at the Jardin, but by the end of the decade, the fashionable world had found new diversions. The director gave up on attracting those who visited the opera, and sought ways to attract and entertain "the people with small purses . . . the public of workers."[27] Although Albert clung to the utilitarian and scientific mission of the zoo, in desperation he began a series of ethnographic exhibitions in which caravans of Nubians, Eskimos, and Argentine gauchos were followed by a group of dwarfs touring Europe as the Kingdom of Lilliput. But Parisians proved to be a fickle lot, and Albert spent lavish sums on concerts to build a clientele during the slow winter season. Albert's activities drew much criticism, especially from the financial reviews, which made science the whipping boy for the Jardin's financial woes. One critic wrote that the Jardin should leave all science to the museum.[28]

While popular ethnography was more scientific than the Jardin's other entertainments, such as puppetry and light opera, it was only marginally profitable, and other Paris institutions were conducting similar programs. With the Jardin on the edge of bankruptcy in 1893, Albert resigned the directorship.[29] Subsequent directors paid only lip service to the scientific goals of the Jardin. The institution achieved its destiny as a hurdy-gurdy fair that continues to delight children, hover near bankruptcy, and belie the intentions of its founders.[30]

The cases of the menagerie and Jardin are instructive, for they illustrate an ongoing tension common to many zoological institutions—the play between serving the general public and also maintaining a collection for scientific research. The museum professors struggled with this tension at their meetings, and although they could not resolve it, they suggested segregating research collections from the public galleries. As we have seen, Isidore Geoffroy Saint-Hilaire wanted separate zoos for practical and pure scientific research. He also proposed closing the museum menagerie to the public for several days per month and limiting the number of visitors. Yet segregation of the public and scientific functions was not always effective, and sometimes objectives became obscure. In the Jardin's case, it was the commitment to applied zoological science which was neither adopted nor continued by another Paris institution.

The Bois de Vincennes Zoo

I conclude with a few remarks about the third zoo of Paris, the Bois de Vincennes zoo. Although Isidore had planted the seed for a zoo in the Bois de Vincennes, the project ultimately grew out of a successful animal exhibition held in the Bois in conjunction with the 1931 International Colonial Exposition.[31] The exposition's commissioner, the former resident-general of Morocco, Marshal Louis Lyautey, orchestrated an exhibition of colonial images which included a reconstruction of the palace of Angkor-Vat and displays of colonial peoples and animals.[32] Some thirty-three million people visited this exhibition, which underlined the value of the people, flora, and fauna of the French colonies.[33] By this time, France's golden age of zoo management had passed, and exhibition organizers had to contract for the display of animals with the Hagenbeck firm of Hamburg.

In the outpouring of nationalistic sentiment and imperialistic enthusiasm that followed the exhibition, the Société des Amis du Muséum recovered a measure of national pride by purchasing the exhibition animals and swaying opinion in favor of the creation of the Bois de Vincennes zoo. Conceived in the 1930s as a sort of permanent exposition of colonial fauna, it is today the largest zoo in Paris and the flagship of French zoos. Thus it was on the eve of World War II that this extended family of the museum realized Isidore Geoffroy Saint-Hilaire's dream of a third Parisian zoo.

These three institutions still exist today and continue to delight their various publics. Visitors to all three zoos will be well served to recall that the institutional and intellectual genealogies of the museum menagerie, the Jardin d'Acclimation, and the Bois de Vincennes zoo share a common history. Though they are now separate institutions, they all trace their origins and institutional distinctiveness to the natural-historical activities of museum scientists and to the prodigious energy and vision of the Geoffroy Saint-Hilaire clan.

THE ORDER OF NATURE

Constructing the Collections of Victorian Zoos

*F*rom the beginning, the Zoological Society of London aimed high. After all, in 1824, when its founder, Sir Stamford Raffles, returned from building his empire in the East Indies, there was no dearth of places for Londoners to view captive wild animals. The royal collection could be viewed downriver at the Tower of London for either a small fee or a small animal to be fed to the lions, and central London boasted an indoor menagerie at Exeter 'Change in the Strand.[1] Single exotic creatures were frequently exhibited as sideshows at fairs and in the yards of public houses. The early decades of the nineteenth century also saw the establishment of the first sizable traveling menageries. Although the patrons of such displays included serious naturalists (for example, during the 1780s and 1790s, Thomas Bewick found models for many of the illustrations in his *General History of Quadrupeds* in the exhibits that reached his native Newcastle), the number of these enterprises suggests that their primary appeal was to a broad popular audience of curiosity seekers.[2]

Entertainment for the masses, however, was not exactly what Raffles had in mind. His years in the East had been devoted to natural history as well as to colonial administration, and he saw these pursuits as complementary. Over time he had identified new species of plants and animals and amassed a large collection of living and preserved specimens. Unfortunately, they all perished in a fire at sea as he returned to the heart of the empire, and he found little awaiting him there to compensate for his loss.[3] What he missed most acutely in London was a collection of living animals worthy of Britain's preeminence. This lack was especially galling because, as an early prospectus for the Zoological Society claimed, though Great Britain was "richer than any other country in the extent and variety of [its] possessions," it could not rival its neighbors' "magnificent institutions" for the display of exotic creatures from such territories.[4] Apparently, others had felt the same lack, as both serious naturalists and amateurs (the latter often members of the aristocracy and gentry) flocked to Raffles' standard. (This was fortunate, as Raffles died in 1826 soon after launching his campaign.)

After an organizational period of several years, the Regent's Park Zoo opened its gates, albeit exclusively to members and their guests, in 1828.[5]

By itself, this complex and formal process of establishment would have defined the Zoological Society of London as a very different institution than preexisting animal collections. The consequences of its unusually elaborate governance were visible in its displays. The composition of most early-nineteenth-century menageries was determined by serendipity—that is to say, some combination of the chances of the marketplace and the public appetite for exotica. Proprietors of what were sometimes called "beast shows" raced one another down the Thames estuary to board ships returning from Africa, the Americas, and especially the East Indies. Such vessels often carried a few living animals along with their more conventional cargo.[6]

Although the Zoological Society was skilled at opportunistic acquisition when the occasion arose, for the most part the animals fulfilled an elaborate master plan. There was always a philosophy of acquisition, although its details varied from time to time, depending upon who was in charge of zoo policy. In fact, even in the years of preparation, when the Zoological Society had no animals and while its founding members were preoccupied with raising funds and securing a suitable site, the question of what the collection should contain provoked sustained debate. There were two major points of view, both of them opposed to what members disparaged as the vulgarity and sensationalism of the displays at such public menageries as Exeter 'Change.

One faction, composed primarily of landowners who projected the zoo as an extension of their own interests in domestic animal breeding, wanted the collection to emphasize species of exotic animals that might be acclimatized for English parks and tables. Various kinds of deer and antelope were frequently mentioned, especially the large and fleshy eland, along with ducks, pheasants, and other fowl and even trout. This group was opposed by the naturalists, who

The crowded indoor menagerie at Exeter 'Change offered the most impressive exotic animal collection in London before the gardens of the Zoological Society of London opened in 1828. (Yale Center for British Art, Paul Mellon Collection)

wished the zoo to stock exotic animals of taxonomic interest, without regard to their attractiveness, edibility, or other usefulness. In 1831 the society's newly constituted Committee of Science and Correspondence, which represented this point of view, compiled a different list that included reptiles, snails, monotremes, and marsupials from Australia.[7]

At first, the Zoological Society tried to satisfy both constituencies. Until 1833 it maintained not only the menagerie in Regent's Park but also a breeding farm at Kingston Hill, not far from London. Stock breeders who wanted to distinguish their livestock with an infusion of exotic blood could pay stud fees that ranged from five shillings for a zebu to one pound for a Brahman bull to two pounds for a zebra. The closing of the farm after a few years reflected both the relatively weak demand for such services, at least when measured in cash rather than rhetoric, and the increasing influence of the scientific wing of the Zoological Society.[8]

Nevertheless, the apparent fact that these two constituencies within the

While Thomas Bewick was preparing the illustrations for his popular *General History of Quadrupeds,* traveling menageries brought both a porcupine and a polar bear to the neighborhood of Newcastle. (By permission of the Houghton Library, Harvard University)

Despite their lofty scientific aspirations, from the beginning the managers of the Regent's Park Zoo were also interested in drawing crowds. (Author's collection)

Zoological Society could not be accommodated under the same roof, or even within the London city limits, may have overstated their incompatibility. It is true that, at the literal level, the competing policies seem quite divergent, if not diametrically opposed. They required different animals for different purposes. The landowners wanted animals relatively similar to those that were indigenous to Britain; the scientists, on the other hand, craved variety of every sort. The landowners wanted animals alive and healthy for breeding; a scientific specimen's best moment was just after it had died, when it could be dissected in the furtherance of comparative anatomy. (The Zoological Society soon established a tradition of generosity with its carcasses, which were all too numerous in its early years; for example, one anatomist had a standing order for the hearts of dead animals, another for diseased joints.)[9]

But however divergent these policies seemed in their treatment of physical animal bodies, whether living or dead, they had more in common when considered as metaphorical expressions. Neither was inconsistent with the rhetorical goal implicitly shared by Raffles and the other founders of the Zoological Society: to acquire, maintain, and display representatives of the animal kingdom in a way that echoed and emphasized British preeminence. At the Kingston Hill farm, this effort to control the natural world took the form of domestication, physical taming, and subordination. (It should be emphasized that the abandonment of the zoo farm did not mean that its goals were abandoned. Rather, for the most part, elite stockbreeders found that they could be more easily and appropriately achieved in other forums, such as those provided by agricultural societies, which routinely considered such matters as hybridization.) In the menagerie itself, the effort to control the natural world was both more abstract and more sweeping. The animal creation was to be not only represented but given its proper designation and put into its proper order. The naturalists who arranged the displays were echoing the work of Adam, if not that of God; the zoo represented the triumph of human reason over the profusion and disorder of nature.[10]

Thus, the Zoological Society of London strove to realize in its collection what one reviewer of zoo guidebooks identified as the goal of the whole science of zoology: "to furnish every possible link in the grand procession of organized life."[11] It pursued as "its principal object . . . to present as many types of form as possible, with the view of illustrating the generic variations of the Animal Kingdom."[12]

The creatures displayed at the London Zoo were conceived as part of an interrelated, graduated zoological series—an animated representation of the standard vertebrate categories. Beginning in 1840, an effort was made to arrange the animals taxonomically. If such arrangements were limited by the availability of housing and by the animals' physiologies, no such constraints operated on acquisitions policy. New animals were often described in terms of their place in the larger order. Thus two jaguars purchased in 1875 were said to have "raised the series of larger members of the cat genus . . . to 20." The kiang, or Tibetan wild horse, donated in 1859, was particularly valued because the Zoological Society had long been anxious to procure "living specimens" so that it could achieve the "object of completing its series of wild species of the genus Equus."[13]

In addition to representing Britain's domination of the natural world, and especially of the exotic places where nature flourished exuberantly, this philosophy served a second purpose. It distinguished the Regent's Park Zoo from other contemporary menageries—none of which could aspire to such numbers of exhibits or to so many animals of scientific or taxonomic interest (that is, animals that would not draw crowds). Thus, in addition to symbolizing Britain as the dominant power in the world, the Zoological Society also constructed its displays to represent the preeminence of social and professional elitism in nineteenth-century British society. It was, however, less successful in this second figurative venture. This was not because Victorian Britain lacked class distinctions; of course, the contrary was true. But one thing that did not divide the classes was the sense of pleasure and participation in British conquest of the globe, whether physical or intellectual. No matter how much the Regent's Park Zoo might surpass other wild animal collections in its financial resources or its scientific sophistication, rhetorically it was *primus inter pares* rather than *sui generis.*

It was therefore impossible for the Zoological Society to distinguish itself from other animal collections. For example, if they could not present taxonomic series that belonged to the entire animal kingdom, smaller menageries often presented those animals in settings that emphasized the triumph of order over the chaotic wild. Dangerous and wide-roaming animals were confined in small cages and placed along well-marked paths, in manicured parks that seemed natural only in contrast to the surrounding urban landscapes. The horticultural displays that routinely adorned the borders, often composed of plants from all over the world, emphasized the artificiality of the setting, as did the constructed lakes, a feature of every zoo that was spacious and prosperous enough to build them. And, again in contrast to the unrestrained profusion of nature, nineteenth-century zoo guides were inveterately linear. No matter what the shape of a zoo, its official guidebook would prescribe a single route through the exhibits, from the entrance to the refreshment stand.[14]

The most powerful visual expression of the human domination of nature was the sight of large carnivores in cages, and as a result, these animals were essential for any zoo of the period. Thus, despite its elevated rhetoric, the Regent's Park Zoo stocked a great many fairly run-of-the-mill big cats. In its early days, their average life span was about two years, which meant that one of them died each month. Their high mortality rate was especially alarming to the Council of the Zoological Society, not on scientific grounds, but because such deaths diminished the zoo's appeal to general visitors, who found the "Carnivora . . . one of the most attractive portions of the Collections."[15]

The rosters of other nineteenth-century menageries were also heavy with large felines. In 1841 visitors to the Liverpool Zoo could admire lions, tigers, lion-tiger hybrids, leopards, and jaguars, as well as a lynx, an ocelot, a margay, and several hyenas (which are not felines but were frequently exhibited with them in Victorian zoos—presumably because they produced a similar effect on observers). A generation later, Bostock and Wombwell's Royal Menagerie was traveling with six to eight leopards, a lynx, approximately ten lions, two tigers, a puma, and a jaguar.[16]

The appeal of these displays was based on the contrast between their natural ferocity and their artificial powerlessness. This was made clear by both the terms in which they were advertised (adjectives such as "fiery" were routine) and the reactions of appreciative onlookers. One visitor to Exeter 'Change observed, "[W]e are . . . surrounded . . . by death under its most frightful form, and yet we hold our life as secure as if we were seated by our own hearths."[17] The sensation was enhanced at mealtimes, which were particularly attractive to visitors. The young Queen Victoria remained behind after one of the performances of Van Amburgh, a celebrated lion tamer, "for the purpose of seeing the animals in their more excited and savage state during the operation of feeding them." So that the queen would not be disappointed, the animals "had been kept purposely without food for six and thirty hours."[18]

Animals could also symbolize British preeminence in ways that called for less aggressiveness on their part or that of their admirers. An elephant named Chunee, who lived for many years in the Exeter 'Change menagerie, may have been the first of the zoo pets (animals widely known and cherished as individuals rather than simply as representatives of their species). Chunee was condemned to death in 1826 because he had become increasingly uncontrollable during what were diagnosed as periodic attacks of sexual excitability, which posed serious

Sir Edwin Landseer captured the essence of the lion tamer's appeal in *Portrait of Mr. Van Amburgh as he Appeared with his Animals at the London Theatres* (1847). (Yale Center for British Art, Paul Mellon Collection)

potential problems in a crowded menagerie that was on the second floor of a commercial building in central London. Because he refused poison, he had to endure a protracted and painful execution by firing squad. His agony inspired unprecedented national attention and an outpouring of public grief. The *Times* printed letters criticizing the fatal decision as well as the manner in which the elephant had been confined during his lifetime; magazines featured both sentimental and humorous poetry on the subject; prints and broadsides portrayed the gruesome details of the killing; and a play called *Chuneelah; or, The Death of the Elephant at Exeter 'Change* enjoyed a successful run at Sadler's Wells.[19]

A violent end was not necessary to inspire compassion in the hearts of the British people; most of Chunee's successors died in their beds. (Indeed, such animals were likely to enjoy favored treatment, since they were powerful draws.) They included Obaysch, touted as the first live hippopotamus to visit Europe since the days of the Roman Empire, who arrived in London with a portable bath and a retinue of native human attendants; Jumbo, the elephant who was sold to P. T. Barnum in 1882 to howls of patriotic outrage; and a succession of gifted chimpanzees, about one of whom a journalist claimed that "paragraphs and articles relating to . . . Sally . . . would fill many goodly volumes."[20] But not all animals were equally eligible. Neither ruminants (giraffes, for example) nor carnivores (including lions and bears) were appropriate candidates. To win the affection of the British public, a wild animal had to be impressive, whether in size, such as the elephant or hippopotamus, or in mental power, such as the chimpanzee, but it could not seem too dangerous or independent.

Zoo pets represented not Britain but their native territories. These were invariably British colonies in Africa and Asia, and never colonies which, like Canada and Australia, had significant European populations. The zoo pets

Broadsides such as this one were among the many popular expressions of grief and interest after Chunee's violent death in 1826. (Bodleian Library, Oxford, U.K.)

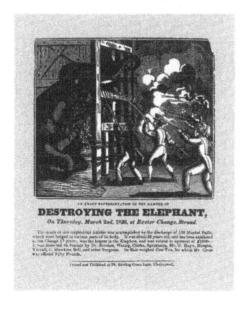

frequently resided in the Regent's Park Zoo, not only because it was the largest and most publicized institution in nineteenth-century Britain but because over the years it became part of the national rhetoric of imperial dominion. When King William IV donated the royal menagerie, which had previously been housed in the Tower of London, to the Zoological Society in 1831, he identified the zoo as the appropriate repository of symbolic acknowledgments of Britain's international position. Queen Victoria routinely consigned to the Zoological Society "the stream of barbaric offerings in the shape of lions, tigers, leopards, etc., which [was] continually flowing from tropical princes."[21]

Although the Regent's Park Zoo had been established to represent an elitism that comprehended both Britain's position in the world and the ascendancy of privileged classes within Britain, the popular appeal of zoological imperialism recast its exhibits as occasions for patriotic, even jingoistic, unity. Rather than representing a unique alternative to other wild animal collections, it offered a more concentrated and forceful version of their symbolism of domination. If the royal family considered the London Zoo a metaphorical extension of its private domains, so did many ordinary visitors to Regent's Park. Especially after the governing council abolished admission restrictions in 1846, the zoo emerged as a national institution, which reflected its glory on all British citizens. And this glory was not solely political or military, whether represented by animals that had been captured in the field, those that had been presented to the queen as tribute, or those that had been adopted as mascots by regiments posted to exotic stations, then relinquished when they grew unmanageable. In a more pragmatic sense, the zoo illustrated Britain's economic prowess; the variety of the animals displayed testified to the range of British commerce. Even the scientific side of the zoo, originally the unstated center of its elitism, could be reinterpreted in the service of popular national pride. Any Briton could take pride in the superior competence of fellow citizens able to maintain so many exotic species in confinement and to manipulate and study them, so that they were better understood and appreciated than by the peoples who had lived among them for millennia.

In summary, keeping exotic animals in captivity was a compelling symbol of human power in general and, depending upon where the animals came from and where they were kept, a symbol of British power. Transporting them safely to Great Britain and keeping them alive were viewed as triumphs of human skill and intelligence over the contrary dictates of nature. Access to the animals' native territories symbolized British power and prestige. The confined and captured animals in Victorian zoos and menageries shed glory on their appropriators and conquerors. The displays in which they were exhibited allowed visitors to bask to the utmost in the reflection of that glory.

A TALE OF TWO ZOOS
The Hamburg Zoological Garden
and Carl Hagenbeck's Tierpark

*T*oday, a few cities around the world can boast of two or even more zoological gardens: for example, New York, Chicago, Tokyo, and Berlin. A century ago, however, just two cities had more than one zoo, Paris and Hamburg, and each zoo was representative of a different process in the development of modern zoological gardens. Of the two French institutions, one, the Ménagerie du Jardin des Plantes—even then almost a century old—was a small and crowded museum of living animals appropriately attached to the Muséum National d'Histoire Naturelle. The other, the Jardin Zoologique d'Acclimatation, was devoted, as the name implied, to promoting the acclimation of exotic animals in France.[1] The two Hamburg menageries were equally different. The zoo in Hamburg which people visit today, Carl Hagenbeck's Tierpark, was only established at its present site in the early 1900s; it is the immediate successor to a line of family-owned Hagenbeck menageries going back to the 1850s.

Purists might suggest that Carl Hagenbeck's Tierpark of a century ago was only a pretentious name coined for the entrepôt of a wild animal dealer and not a zoo at all. Nevertheless, *Tierpark* has since become the German term for a zoological park. In the modern sense of the term, it was a genuine zoo, albeit a small one in area and with a largely transient animal population. Yet by the turn of the century, the value of the Tierpark's animals was "greater than the value of the animals in any one zoological garden in Europe."[2] The "official" Hamburg Zoo one could visit a century ago was different from any of the Hagenbeck zoos, new or old; it was a classic example of a large and ambitious society-run garden, directed by scientists, manipulated by patrons, and largely ignored by City Hall. The zoo's thirty-five acres on a site just behind the old city walls were leased from Hamburg's Senate (city government) free of charge—but only for fifty years (later the leases were renewed every ten years).[3] At no time did the zoo receive subsidies; but then, for most of its history it required none.[4]

The Hamburg Zoological Garden, 1860–1920

While the Hagenbecks were developing their animal trade, the Zoological Society of Hamburg was organized in 1860 by a group of wealthy local merchants and public servants for the specific purpose of establishing a zoological garden "for the study of nature, especially that of animals, and for the recreation and education of the people."[5] Initially, eight hundred shareholders contributed the capital needed to lay out a zoological garden. It was largely completed by November 1862 and was conveniently located adjacent to a major railroad station.[6]

It was only after the new zoo was ready for opening that the society's council decided on a director. The choice fell on Alfred Brehm, a name not well known, if at all, outside central Europe. But in German-speaking states Brehm's name is synonymous with nontechnical reference books on animals. What, at first, recommended him to the society's council was his experience in collecting animals and managing menageries while on scientific expeditions to Africa and Spain.[7] As zoo director, Brehm found the time to write the first volumes of the popular zoological encyclopedia *Illustrirtes Thierleben,* which was to make his a household name. But it was time spent which cost him his office only three and a half years later.[8]

The Hamburg Zoological Garden opened its gates to the public for the first time on May 17, 1863; 1,839 burghers came. On August 2, the garden had a peak of 38,137 visitors; on December 5, only 1. Altogether, 225,553 came in the first six months. In its best years it would attract 600,000—good attendance in an age when zoos could be particular about whom they let in.[9]

The new zoo in Hamburg was only the seventh in Germany. Within five months of opening, it boasted 300 species. For the next twenty years its animal

The waterfall connecting two waterfowl ponds was a landmark of the Hamburg Zoological Garden. The "Egyptian" stork house at the right was one of only two animal houses with exotic architecture the zoo would ever have. The bear tower at the left was typical of the heavy, Germanic style favored for bears in zoos from Alsace to East Prussia. (Lithograph courtesy of W. Kourist)

collection would be the most varied in the nation, later surpassed only by the Berlin Zoo. In its most successful years, those immediately before the First World War, the collection counted more than 4,000 specimens representing 1,000 species. Even in its last full year of existence (1929), the garden displayed 880 species.[10]

The site of the zoo was hilly with initially low trees; the two highest points were crowned by a chamois hut and an owl house with a lookout tower, respectively. Three large ponds were connected to one another by running brooks; a natural slope created a small waterfall. A porpoise acquired in 1864 from the Hamburg fishmonger–cum–animal dealer Carl Hagenbeck Sr. was let out in one of the ponds—the second cetacean ever in a zoo.[11]

Visitors coming through the main gate first came across the deer paddocks, always among the most interesting in a European garden. During its first decade, the zoo exhibited eleven passenger pigeons and thirty-nine Carolina parakeets. The year before the very last Carolina parakeet died in the Cincinnati Zoo (1918), the last of the species in Europe expired in Hamburg. The garden had the rather awkward distinction of being the last to have a Cape lion (1888) and a Burchell's zebra (1915) before those two prominent subspecies became extinct. But it was also the first in Europe to exhibit the now extinct Schomburgk's deer (1862), a Rocky Mountain goat (1880), a Himalayan blue sheep (1882), an African forest elephant (1882), and a zebra duiker (1903), to name only a few big-game animals.[12] The first tapirs born in a zoo (a Brazilian in 1868 and a Malayan in 1879) were "Hamburgers."[13] Like other large zoos of its age, the Hamburg Garden was keen to collect zoological treasures—all to top the perceived competition.

Germany's first aquarium building opened at the Hamburg Zoo on April 26, 1864. The pachyderm house of 1867 was replaced by a new one in 1881, housing the three most popular animals the zoo would ever have: the Indian bull elephant Anton (a magnificent tusker), the Indian bull rhinoceros Begum, and the breeding bull hippopotamus Bachit. All three lived for thirty years, more or less simultaneously, at the zoo.[14]

Only two of the animal houses were built in the exotic style popular among the larger Continental gardens: the "Egyptian" stork house and the new ostrich house of 1904, covered with "Cape Dutch" tiles. Germany's only marsupial house was built in 1895, its only house for native birds—easily the most popular avian exhibit—in 1897.[15]

The original primate house was replaced in 1915 by the largest ever built up to that time. War, unfortunately, brought famine and influenza to *all* primates in Germany; it was years before the sixty-nine inside and twenty-two outside cages could be properly occupied. The Hamburg Zoo was the only one in Germany able to maintain any anthropoid apes—two chimpanzees—through the First World War. The Amazonian manatee, too, survived to establish a longevity record for sirenians of thirteen years.[16]

Hagenbeck's Tierpark, 1850–1910

The letterhead of the Hagenbecks' zoo gives a foundation date of 1841. Family legend has it that the animal trade began with six harbor seals caught in a fisherman's net and dumped with the rest of the take upon the doorsteps of the Hamburg fishmonger Carl Hagenbeck Sr. Truth or not, Hagenbeck certainly maintained a small animal dealership—a pet store, really—along with his wholesale seafood business through the 1850s.[17]

The animal trade picked up in the early 1860s, when something of a boom set in, as zoological gardens became established in central Europe at the rate of almost one a year. Hagenbeck's dealership became wholly independent of the seafood store in 1863. On Spielbudenplatz (Gaming Booth Square) in the heart of what was then, as now, Hamburg's red-light district, C. Hagenbeck's Handels-

Carl Hagenbeck Sr.'s wholesale seafood market and menagerie was the birthplace of Carl Hagenbeck Jr., founder of the Tierpark. The house still stands and now serves as a residential building. The elder Hagenbeck is at the left, holding a pole to a sturgeon. (Reproduction of undated painting by Johannes Gehrts, author's collection)

Carl Hagenbeck's Handels-Menagerie was established near Hamburg's harbor in 1863. Seamen, like the one at center, were important sources of wild animals to dealers in the nineteenth century. Carl Hagenbeck Jr. is shown at left, taking a monkey out of a crate. (Reproduction of undated painting by Johannes Gehrts, author's collection)

Menagerie opened to the public in the same year, incidentally, as the Hamburg Zoological Garden.

The Handels-Menagerie ("dealer's menagerie") was actually an entrepôt, not a zoo; two street-front shop galleries sold monkeys and parrots respectively. In a courtyard behind the shops, bird cages were stacked between large tubs with seals. A barnlike structure eighty feet by thirty feet had stalls for carnivores on one side, herbivores on the other; crates of boas and pythons lined the middle. In a second courtyard beyond a back street, more birds and exotic domestic animals were groomed for sale. Yet here was Europe's first African rhinoceros on exhibit since Roman times, and its first Sumatran rhinoceros.[18]

In 1866 Hagenbeck Sr. passed on ownership of the Handels-Menagerie to his eldest son, Carl Jr. The young Hagenbeck expanded the animal dealership with such success that within a decade it was the largest in Europe. In 1874 the junior Hagenbeck moved the business from its crowded Spielbudenplatz quarters to Neuer Pferdemarkt (New Horse Market), about a mile and a half north, and opened his first "Tierpark," an entrepôt-cum-zoo of two acres replete with lion house, elephant house, monkey house, reptile house, birds-of-prey aviary— in fact, most everything any other zoo would have, only more crowded still and more valuable than most.[19] The Neuer Pferdemarkt zoo was the first in Europe to exhibit such rare mammals as the Somali wild ass (1882), the gerenuk (1883), the pygmy hippopotamus (1884), the African manatee (1887), the Mongolian wild ass (1900), and Przewalski's wild horse (1901).[20]

What Hagenbeck liked to call anthropological-zoological exhibitions were introduced in 1874: Lapps accompanied a shipment of reindeer and put on a show reenacting daily life in Lapland in front of enthusiastic audiences in Hamburg, Berlin, and Leipzig. Over the next fifty-five years, Hagenbeck's Tierpark would organize some seventy performing ethnographic shows with groups rang-

Carl Hagenbeck Jr. (1844–1913) revolutionized animal keeping with the development of the "panorama," a bar-less exhibit laid out in successively higher stages containing different species, separated from one another and from the public by moats. (From a 1913 guidebook to Carl Hagenbeck's Tierpark, Stellingen, Smithsonian Institution Library, National Zoological Park Branch)

ing in size from three to four hundred, featuring three dozen tribes and "races." During the 1880s, some shows attracted almost a hundred thousand spectators a day. The zoo on Neuer Pferdemarkt was too small itself to stage very large performances. As a result, some performances were held in the Hamburg Zoological Garden, others on the spacious grounds of Hagenbeck's brother-in-law J. F. G. Umlauff, a dealer in natural history objects.[21]

Apartment houses surrounded the Neuer Pferdemarkt Tierpark in what became a densely populated neighborhood. The roars and growls of carnivores, such as these bound for export, added to the traffic noise the Hagenbecks and their neighbors had to bear. (Undated photograph, photographer unknown, published in an album by Carl Hagenbeck; author's collection)

Elephants were regularly exercised around the common of Hagenbeck's old Tierpark on Neuer Pferdemarkt. The lion house at the left housed Europe's first Persian leopard and second Siberian tiger. (Undated photograph, photographer unknown, from the Hagenbeck archives)

This illustration, from the 1913 Stellingen Tierpark guidebook, provides an example of the diversity of Hagenbeck's ethnographic exhibits. (Smithsonian Institution Library, National Zoological Park Branch)

"Ethiopia" was the title of a 1909 ethnographic show at Hagenbeck's Tierpark in Stellingen. A re-created "Ethiopian" village served as a set for this show. (Author's collection)

Hagenbeck's Dressurhalle (center for training animals) at the Neuer Pferdemarkt Tierpark was unique among zoo buildings of the last century. It was here that non-violent methods to train animals became standard. Training demonstrations of polar bears and big cats in a single enclosure helped to prove Hagenbeck's point (photo ca. 1890). (Unknown photographer, from the Hagenbeck archives)

Wanting to make good use of Germany's African colonies, Carl Hagenbeck tried to harness the zebra. Lorenz Hagenbeck (*left*) and the animal collector Jürgen Johannsen (*center*) managed to get zebras to pull them around the Neuer Pferdemarkt Tierpark, but little else (photo ca. 1900). (Unknown photographer, from the Hagenbeck archives)

One of the buildings on Neuer Pferdemarkt which other zoos did not have was a "school" to tame and train animals. Tame animals, of course, and even more so performing animals, brought better prices than truly wild beasts. Genuinely disturbed by the harsh, even cruel, methods standard among the trainers a century ago, Hagenbeck introduced what he termed *zahme Dressur* (roughly: training without the use of force or intimidation). His trainers were instructed to recognize the individual intelligence and nature of each of their charges, coaxing and encouraging them rather than thrashing them with canes or poking them with hot irons.[22] The enormous success of the various acts of performing animals in his circus, established in 1887, proved his point: Hagenbeck's animal acts quickly became famous—they were featured at the World's Fair in Chicago (1893) and Saint Louis (1904).[23]

The old Tierpark was to flourish for thirty years. As a zoo, entrepôt, and circus headquarters, however, two acres were really too small, despite lots

The year Carl Hagenbeck Jr. acquired his patent on the "panorama" (1896), he staged a polar exhibit on Hamburg's Heiligengeistfeld circus and fair grounds to introduce his favorite "invention" to the public. The moat separating seals from polar bears was not visible. (Illustration published in the Leipziger *Illustrierte Zeitung* of January 7, 1897)

The Arctic panorama of Hagenbeck's new Tierpark could originally be viewed both from above and from the main path. Wartime bombing destroyed the huge "rock" topping the panorama, and the reindeer paddock to the left, in 1943 (photo ca. 1908). (Unknown photographer, published in an album by Carl Hagenbeck; author's collection)

purchased and rented elsewhere in the city. After years of looking in vain for a suitable site in Hamburg, Hagenbeck acquired at the turn of the century an estate in Stellingen, then a suburb of Hamburg. There he could build a true zoological park, encompassing all of the principles of keeping wild animals in captivity which he had come to recognize in his decades as an animal dealer, circus proprietor, and zoo director. The widths of the moats, for example, which he substituted for bars and fences were determined via the circus ring by observing the distances that animals could jump. None of his principles was really new, but they had yet to be put into practice in a zoological park.

In 1896 Hagenbeck received a patent for what he called the "panorama": a series of enclosures laid out as stages, one behind and slightly higher than the other, separated from one another and from the public by concealed moats. The animal houses and the walkways between enclosures were screened by artificial rock work and hedges. Within five years, twenty-five acres of flat potato fields

Carl Hagenbeck Jr.'s new twenty-five-acre Tierpark opened in May 1907. The main attractions were the Africa panorama (the row of enclosures in the foreground) and the Arctic panorama (in the upper-left corner). The monumental building in the upper-right corner was the animal trade depot, destroyed by bombs during World War II (view drawn for Carl Hagenbeck in 1907 by August Urban; author's collection)

Sculptures by Joseph Pallenberg adorn the main entrance to Hagenbeck's Tierpark. The Sioux and Somali warriors that flank the gate remind visitors of the ethnographic shows that were once part of a day at the zoo. (Postcard, Smithsonian Institution Library, National Zoological Park Branch)

with six trees were transformed into a landscape of mountains, gorges, lakes, and islands which initially comprised two panoramas: Africa and the Arctic.[24] The exhibits created a huge sensation on opening day, May 7, 1907. Everyone attending (more than ten thousand visitors) seemed to experience the chilling sensation that lions and polar bears could leap from their exhibits right into the public areas.[25]

Another thirty acres were developed by 1909, including Germany's first ostrich farm and a dinosaur park of life-sized sculptures.[26] An arena for ethnographic shows provided Hagenbeck the opportunity to organize large performances for the first time on his own premises. The most popular was a Wild West show in 1910, with ten cowboys and forty-two Sioux Indians from South Dakota's Pine Ridge Reservation. More than a million spectators streamed into the Tierpark that summer alone.[27] Most of the implements and works of art

This 1910 American Wild West show was the most popular ethnographic show that Hagenbeck's Tierpark was ever to mount. More than one million visitors saw these Oglala Sioux dance at the foot of an artificial mountain. (Author's collection)

Since the turn of the century, Carl Hagenbeck Jr. (*center*) received active support from his sons and subsequent heirs, Heinrich (*left*) and Lorenz. Heinrich was largely responsible for zoo operations, Lorenz for the circus. The two-and-a-half-year-old Bengal tiger was a born "Hagenbeck" (photo ca. 1908). (Unknown photographer, published in an album by Carl Hagenbeck; author's collection)

collected for a show were later donated to the Hamburg Museum of Ethnography, if not sold at the Tierpark's souvenir bazaar.

Eighty or ninety years ago, most zoological gardens were acquiring their stock from animal dealers. Hagenbeck's dealership, the world's oldest and largest, was right there in a Hamburg suburb, but that was not where the "official" Hamburg Zoo was buying its animals any longer. As a matter of fact, as Bronx Zoo director William Hornaday noted in the *Bulletin of the New York Zoological Society,* "the zoological garden directors of all Germany were industriously engaged in boycotting Mr. Hagenbeck . . . because [he] had had the temerity to build at Hamburg a private zoological garden so spectacular and attractive that it made the old Hamburg Zoo look obsolete and uninteresting."[28]

What is now taken for granted by almost every visitor to a zoo—moated exhibits in a landscape simulating nature; gregarious animals of mixed species kept in herds in large enclosures; and animal performances based on conditioning and sensitivity, not on brute force and intimidation—all started at Hagenbeck's Tierpark.

CARL HAGENBECK JR. DIED IN 1913 and was succeeded as proprietor by his two sons, Heinrich and Lorenz. Lorenz kept the firm solvent through the First World War, heading the circus through the neutral states of Europe. Neither Hagenbeck's nor the society's zoo was making a profit, however, in the wake of Germany's economic collapse following the 1919 Treaty of Versailles. On October 3, 1920, the Tierpark closed down for three and a half years, until the nation's economic climate warmed up, and Lorenz's son Carl-Lorenz was able to reestablish the animal trade.[29]

The Zoological Society of Hamburg went into liquidation at the end of 1920, and the Zoological Garden was scheduled for closure the following January. A group of patrons succeeded in initiating a new company, the Hamburg Zoological Garden Corporation, to take over the lease. Thus the zoo managed to stay open—if only for another decade. During this tenuous period, the Hamburg Zoological Garden was superintended by Julius Vosseler, only the garden's third director in sixty years. Despite Vosseler's efforts, the Hamburg Zoological Garden was dissolved in 1930 at the start of the Great Depression.[30]

The Zoo Corporation decided to divide its lease into a bird park and an amusement park. The bird park incorporated about a third of the old zoo site, including the large ponds for waterfowl and the bird houses, of course. The Hamburg Bird Park (in German: Hamburger Vogelpark) opened its gates— a side entrance of the old zoo—on July 7, 1930. Only twelve months later, however, the Zoo Corporation went into liquidation, and the bird collection was sold to the Fockelmann pet shops in Hamburg. Hamburg became a single-zoo city.[31]

The Hamburg Zoo of today, Carl Hagenbeck's Tierpark, remains a pleasant and, in parts, even a uniquely beautiful zoological park. But like the Hamburg Zoological Garden of a century ago, or the Tierpark on Neuer Pferdemarkt, it is now, by the standards of its time, very much a conventional zoo. The Hagenbecks can claim to have once set the avant-garde standards in zoo design, but that is now history.

ZOOS AND AQUARIUMS
OF BERLIN

*A*lthough nineteenth- and twentieth-century documents containing information on zoos and aquariumlike institutions in or near Berlin refer to five facilities, my emphasis is on the evolution of three of them: the Peacock Island Menagerie (Menagerie auf der Pfaueninsel), the zoo near Berlin (Zoologischer Garten bei Berlin) which eventually became the Berlin Zoo, and the Aquarium Unter den Linden. Their histories are very much connected.

Peacock Island Menagerie (Menagerie auf der Pfaueninsel)

The creation of a menagerie by Prussian king Friedrich Wilhelm III on Peacock Island (Pfaueninsel), situated in the Havel River near the royal palace at Potsdam (about fifteen and a half miles from the heart of Berlin), was a first step in the establishment of a modern zoo in Berlin. Originally designed by Johann Eyserbeck in 1795, Peacock Island's landscape was changed to what was then called an "ornamental" farm in 1805 by Johann Fintelmann, the royal gardener.[1] With Fintelmann's help, King Friedrich Wilhelm III collected exotic animals such as peacocks, birds of prey, and monkeys for the menagerie.

By 1822 the famous landscape gardener Peter Lenné had begun his creative work on Peacock Island. His intention was to create a variety of floral and faunal scenarios, with all the animal houses, enclosures, and cages integrated into the overall design.[2] In conjunction with the famous architect Martin Rabe, Lenné worked out a new concept that relocated the menagerie to the center of the island. Eight animal houses were designed and added to the deer and buffalo enclosures originally built in 1802. A mechanical building to operate a fountain was built, and pheasants from the nearby New Park of Sans Souci were brought to Peacock Island.[3]

In 1824 Friedrich Wilhelm III agreed to a modest expansion. Rabe erected a waterfowl house and a bear den, and in 1828 the architect Albert Schadow built thirteen enclosures for a deer park, cages for foxes and wolves, a house for boars and pigs, and a house for kangaroos. Schadow had also supervised the completion of a llama and antelope house, a beaver valley exhibit, and an enclosure with a creek for buffalo by 1834.[4]

The Peacock Island Menagerie was something of a missing link between a royal menagerie such as Austria's Schönbrunn and a state-of-the-art zoo of the nineteenth century. Like Schönbrunn, it was open to the public only on certain days of the week.[5] Its animal buildings and enclosures were not arranged around a central pavilion area but were located along different landscaped pathways. Besides its antiquated, menagerielike facilities, it had some modern aspects. For example, each indoor cage in the monkey house had a separate outdoor enclosure. The Paris Zoo of the nineteenth century was the first to build a monkey house, with a single outdoor enclosure connected to several small indoor cages; thus the visitor was able to observe the animals outdoors or indoors. Generally, however, it was not until the end of the nineteenth century that zoos began adding outdoor enclosures for monkeys.

Like all menageries, the Peacock Island Menagerie was a feudalistic institution, integrated into the park that surrounded the palace in Potsdam. History shows that its existence was dependent solely on Friedrich Wilhelm III's interest in exotic wildlife. In 1832 the menagerie housed 847 specimens representing 96 species. Ten years later, in 1842, it housed only 65 species with 507 specimens, a decline that had begun after the death of King Friedrich Wilhelm III in 1840.[6]

King Friedrich Wilhelm III's successor, Friedrich Wilhelm IV, was not an exotic animal enthusiast. In 1842 the king gave away the animals and even some of the buildings (which were relocated) to a new, independent zoological garden associated with the Berlin University and Museum. Subsequently, the menagerie closed down. Today, only the bird house and the bear den remain on the island. Of the two, only the bird house is open to the public.[7]

The Zoological Garden near Berlin (Zoologischer Garten bei Berlin)

The zoo near Berlin owed its existence to the initiative of Professor Martin Lichtenstein and to the donations of Friedrich Wilhelm IV. Interestingly, the same people who were responsible for planning and supervising the Peacock Island Menagerie were now appointed to design and manage the new zoo: Professor Lichtenstein of the Berlin University and Museum; Peter Lenné; and August Siebers, the supervisor of the earlier menagerie.[8]

Since visiting London in 1832, Lichtenstein wished to found a zoo independent of the king. In 1840 he wrote his essay "Thoughts on the Installation of Zoological Gardens near Berlin" ("Gedanken über die Errichtung Zoologischer Gärten bei Berlin") and obtained the approval of Friedrich Wilhelm IV. On January 31, 1841, the king gave an order for the creation of a committee, with Lichtenstein, a certain Minister von Ladenburg, Alexander von Humboldt, and Lenné as members, to work out the details.[9]

Peter Lenné, designer of the new zoo, had long supported Lichtenstein's goal. Lenné had worked out a design concept for the zoo as early as 1833,[10] but it never got the support of Friedrich Wilhelm III. After Lichtenstein's return from London, Lenné immediately developed designs for a section of the Tiergarten ("animal garden"), the former hunting grounds of the kings, which are today a public park for the people of Berlin. One of Lenné's many designs was implemented, and the zoo was opened to the public in 1844. The Berlin Zoo is still located there.[11]

Lichtenstein and the other members of the committee agreed that a private zoo could only survive as a stock corporation. This corporation was legalized on February 27, 1845. Strangely, Friedrich Wilhelm IV gave his assent to the corporation on May 7, 1845—nearly nine months after the opening of the zoo. In any event, the corporation was not very successful; by 1869 it had sold only 191 shares.[12]

The zoo was originally designed to be a park with only a few buildings, enclosures, and animal cages, resembling the menagerie on Peacock Island. Two small ponds were built, but all trees remained untouched (a prohibition against cutting trees existed there until 1869).[13] At its opening the zoo had only a few animal houses: two older facilities from the pheasantry that existed in the park prior to the zoo's creation, plus the "bear castle," bear house, monkey house, bird house, buffalo house, and a restaurant for zoo visitors. Between 1853 and 1860, new construction included two carnivore houses, an elephant house, and a domesticated-fowl house. By adding smaller-sized cages and enclosures, many more animals could be exhibited than on the zoo's opening day.[14]

Building activities did not change the overall character of the zoo: the ponds, hills, wooded areas, and footpaths were still the same as Lenné had planned. Although attractive to the public, the zoo was not linked to a scientific society as was the Regent's Park Zoo in London, but it retained its connection to Berlin's university and museum scientists. Zoo staff members, university faculty, students, and museum scientists conducted scientific projects involving the animal collection and published papers.[15]

During the 1850s and 1860s several other German cities constructed zoos: Hamburg, Hanover, Cologne, Frankfurt, and Dresden. Some were stock corporations, and others were privately run. Their animal houses were spacious and the animal exhibits large, and their grounds were full of flowers.[16] By the mid-1860s, the Berlin Zoo, now the oldest German zoo, began to look antiquated compared with the designs of the new zoos. Hartwig Peters, of the Berlin Zoological Museum, was given supervision of the zoo after Lichtenstein's death in 1857; he initiated a total reorganization in 1867. Peters, in turn, was succeeded in 1869 by Heinrich Bodinus, former director of the Cologne Zoo. He was the zoo's first scientific director and the first to take full responsibility for all zoo operations. He and his board of directors agreed to a total modernization of the Berlin Zoo.[17] Money was raised by selling stock. Within a very short period of time a significant number of shares were sold. This enabled the zoo to begin a reconstruction phase that would entirely change its appearance.[18]

Three new ponds were built; new buildings and animal enclosures were erected. Pathways leading to the new facilities were also built. The most spectacular change, however, was the biological systematization of the zoo. Bodinus established cattle and deer exhibits, an antelope house, a new carnivore house, a new bear castle, and a house of perissodactyls and pachyderms.[19] He also tried to exhibit as many species of each order and family as possible. During the nineteenth century, this was common practice in all zoos and corresponded to the perspective of scientific inquiry then in vogue.

As director of the Cologne Zoo, Bodinus had become quite interested in exotically designed buildings. He had very likely seen the elephant house in Antwerp, the first to be built like an Egyptian temple.[20] As a result, between 1869 and 1873 the following exotically designed animal houses were built at the Berlin

The first monkey house was opened in 1844. After 1869 it was used to keep birds and rare mammals such as lemurs. It was destroyed in 1943 by the war. The architect, Johann Heinrich Strack, used the monkey house in the Jardin des Plantes in Paris as a model. (Zoo Guide 1864, author's collection)

The first elephant house was built by architect Gustav Herter in 1859. It was planned for only one elephant and zebras and giraffes. It was a castlelike building with some Italian and other exotic elements. (Zoo Guide 1864, author's collection)

Zoo: the "bear and carnivore castle," the "antelope mosque" or "palace," the "pachyderm Indian temple," and the Moorish-style bird house.[21] In all, the zoo had changed from a park with few enclosures and scattered animal houses to an animal collection with spacious animal houses and numerous enclosures, new ponds, and an ever-growing restaurant with band stands.

Bodinus died on November 23, 1884. His successor, Maximilian Schmidt, was a veterinarian and former director of the Frankfurt Zoo. He published the first book on diseases of animals in captivity.[22] During his three years at the Berlin Zoo, he built the first hippopotamus house and reorganized the zoo's administration. The Berlin Zoo exhibited 166 species in 1864, 504 species in 1884, and 544 species in 1888.[23] This increase was the result of the changes that Bodinus had initiated and Maximilian Schmidt continued.

Schmidt's successor was Ludwig Heck, former director of the Cologne Zoo. Heck continued the work of his predecessors, building many new exotically designed animal houses for ostriches, camels, zebras, and wading birds.[24] One of the better known of these buildings was the wading bird facility with its Japanese design. The ostrich house was designed like an Egyptian temple, resembling closely the one built forty-five years earlier in Antwerp. Painted murals and hieroglyphs designed by teachers of the Berlin University embellished the building. This was truly a unique animal house.

At the turn of the century, more and more animal houses were erected, and new pedestrian pathways changed the face of the zoo. It was less of a "garden" and more like a "town." To compensate for the increase in architectural features, Heck designed rustic and naturalistic buildings such as the boar house with its swamp in the front, a goat and sheep rock, a wooden Russian manor house for European bison, and a birds-of-prey rock. During the decades before and after the turn of the century, more species were added to the collection. By the outbreak of World War I, which halted all zoo construction, 1,474 species of

The third monkey house presented a fantastic exotic style. The architects Hermann Ende and Wilhelm Böckmann designed a greenhouse with inside cages. At the opening in 1884, however, it had two outdoor cages. The building was changed to the large "Affenpalmenhaus" in 1924. (Illustration published in the Leipziger *Illustrierte Zeitung* 14 [1885])

The house for wading birds stood from 1897 until 1943, when it was destroyed by the war. The architects Heinrich Kayser and Karl von Großheim designed a Japanese-style building. Like the Egyptian temple, it was an outstanding example of a realistic adaptation of an exotic style. (Zoo archives, author's collection)

The Egyptian temple for ostriches opened its doors in 1901. The architects Heinrich Kayser and Karl von Großheim worked together with the Egyptologist of the university to create realistic Egyptian paintings and hieroglyphs. The ostrich house was destroyed during the war in 1943. (Zoo archives, author's collection)

mammals and birds lived in the zoo, many of them as solitary specimens.[25] Clearly, breeding did not play the major role it does today.

During the early 1900s, Hagenbeck had opened his Tierpark outside Hamburg, in Stellingen, Germany. Its avant-garde, panoramic design—with open, moated enclosures and diverse species from a single geographic zone in one exhibit area—attracted considerable public attention. Heck did not at all agree with Hagenbeck's new ideas and saw the end coming for the systematic zoo in which closely allied taxa were exhibited together despite living on different continents or in different ecosystems. It was not until the 1930s, when Heck was succeeded by his son Lutz, that modernization started and the zoo began to implement Hagenbeck-like exhibits. The Berlin Zoo (formerly the West Berlin Zoo) today is as popular as ever and is one of the world's premier zoos.

Aquarium Unter den Linden and the Berlin Zoo-Aquarium

In the 1860s, interest grew on the part of the Berlin business community for an aquarium that would be a profit-making stock corporation. In 1867 the Commandit-Gesellschaft auf Aktien Berliner Aquarium was established with capital of two hundred thousand thaler.[26] The building site was to be "Unter den Linden" (along a major avenue), in the center of town, not at the zoo. The stock corporation proved to be very successful, paying about 2 to 5 percent in dividends annually. The aquarium opened in 1869.

The aquarium's first director, Alfred Brehm, former director of the Hamburg Zoo from 1863 to 1866, served until 1874. Brehm did not envision the aquarium as a profit-making institution; he wanted it to be of educational value to the public and to provide research opportunities for himself and his colleagues.[27] With this in mind, he published articles in popular journals and wrote popular books such as *Illustriertes Thierleben* (*Illustrated Animal Lives*, 1864–69).

The house for the European bison was built like a Russian wooden manor house. Designed by architects Carl Zaar and Vahl, it was opened in 1905 and still exists. (Zoo archives, author's collection)

With its emphasis on education, the aquarium was designed like a grotto, part of it made of natural rock. The "Geologische Grotte" depicted "the strata of the earth's crust." The grotto also featured birds and pools for seals.[28] The Aquarium Unter den Linden was a three-story building. Machinery and water tanks were on the ground floor, aquarium basins for the fish on the first floor. Because of Brehm's special interest in birds,[29] a huge aviary, with cages for mammals placed around it, was located on the second floor. From the 1860s until the closing of the aquarium in 1910, few changes were made in the facility.

In 1874 Otto Hermes, vice-director, took over. He was a pharmacist and scientist and politically active, as a member of parliament. Hermes conducted research at the aquarium and wrote both scientific and popular papers. His work at the aquarium contributed a great deal to the improvement of aquatic animal keeping.[30] Hermes succeeded in producing artificial salt water that could sustain oceanic fish. The salt water production was expanded; Hermes sold it to other

The architect Wilhelm Lüer from Hanover built the Berlin Aquarium following the ideas of A. E. Brehm. The aquarium in the first floor was designed like a natural grotto. Brehm and Lüer used natural rocks from different parts of Germany to show the geological variety of the earth's crust. (Illustration published in the Leipziger *Illustrierte Zeitung* 52 [1869])

The large aviary of the Berlin Aquarium was centrally located on the second floor. It was divided and fronted with vertical piano wire, which was nearly invisible. (Illustration published in the Leipziger *Illustrierte Zeitung* 52 [1869])

German aquariums such as those in Munich, Hanover, and Dresden. He also found better ways to transport marine life by rail from the Mediterranean Sea. He imported a number of animals and traded many of them with other institutions.

The trading place for Mediterranean marine life was at Trieste in Italy. Hermes sent one of his employees to set up a trading post in nearby Rovigno, Croatia, which proved to be very successful. Besides its function as a trading operation, the post had a laboratory for scientific studies which became a well-known biological research station that is still in use. The post was run under the guidance of the Aquarium Unter den Linden.[31]

The Berlin Aquarium exhibited many species of general interest. Hermes, however, was especially interested in apes, and gorillas in particular. He housed one or more apes at a time in a cage near the aquarium's aviary.[32] In 1876 Hermes exhibited the first gorilla in Germany (the second to be exhibited in Europe).[33] Later, gorillas were exhibited intermittently from 1881 through 1907.[34]

Other mammals of great interest exhibited at the aquarium were the walrus (1884 and 1886), North American manatee, and a gray seal between 1887 and 1905. Also displayed were reptiles, such as the Chinese alligator (the first in Europe), and interesting birds, such as passenger pigeons, Carolina parakeets, and roseate spoonbills.[35]

Relations between the zoo and aquarium were good. In the 1860s and 1870s aquarium director Brehm and the zoo's director, Bodinus, were friends and worked together closely. After the aquarium opened its doors in 1869, the zoo gave up exhibiting fishes and reptiles. It was not until 1891 that fish tanks for exhibiting aquatic life were installed in the zoo's antelope house.[36]

With the beginning of the twentieth century, prices for real estate near and in the city skyrocketed. Wealthy financial groups were able to gain control of the aquarium's land. Baubank-Union bought enough shares to close down the aquarium.[37] Hermes tried to find a different location for a new aquarium and even negotiated with the zoo's director at that time, Ludwig Heck. Hermes died on March 19, 1910, however, before negotiations were completed. On October 30, 1910, the Aquarium Unter den Linden closed its doors, and all the animals were sold to facilities in Frankfurt and Leipzig.[38] In 1913 a new aquarium opened on the zoo premises. It was planned and supervised by Oskar Heinroth, who later became its official curator and director.

The Berlin Aquarium of today, a part of the Berlin Zoo, places its emphasis on thematic fish exhibits. It has been enlarged and modernized over the years and now emphasizes the breeding of rare and endangered species, especially disappearing coral reef fishes. The breeding of selected terrestrial endangered species such as the Komodo dragon and the Muroroa snail has also become a significant function of the aquarium.

THIS COMPILATION OF THE BRIEF histories of the Peacock Island Menagerie, the Berlin Zoological Garden, and the Aquarium Unter den Linden serves to illustrate the complex evolution that many zoos and aquariums experienced in the nineteenth century. In fact, the process fits the pattern outlined by Michael Robinson in his foreword to this volume. First, there was a royal menagerie established during the period of imperialistic expansion at the end of the eighteenth century. In the nineteenth century, the menagerie became independent of the crown, managed by a university professor with natural history museum affiliations. Along with a board of directors, he developed a "classical" zoo open to the public on a regular basis. The aquarium developed in the late 1800s when glass technology was sufficiently advanced. The first aquarium, a stock corporation built in 1869, closed in the early 1900s. The Berlin Zoo then built its own aquarium and made it a part of the zoo. A process not unlike this occurred among zoos and aquariums in many large cities around the world in the nineteenth and early twentieth centuries.

A PARADOX OF PURPOSES
Acclimatization Origins of the Melbourne Zoo

*S*ome institutions have short, predictable origins. Others have protracted and paradoxical ones. Zoos reflect both. Other chapters in this volume describe the recreational and scientific origins of European zoos. This chapter describes the very different origins of Australia's first major zoo. Although both recreational and scientific elements influenced its birth, another grand desire was preeminent: the desire to improve the Australian colonies both materially and aesthetically by the introduction of new plants and animals. The mid-nineteenth-century progenitor of Australia's first zoo functioned as a depot for introduced animals awaiting distribution. Only later was it transformed into a zoological garden that would entertain and educate the public.

In the mid-nineteenth century, the young British colonies in Australia lacked resident royalty and an ensconced elite to develop private menageries and support the formation of public zoological gardens. There was no Charles I requiring the stocking of royal game reserves or a Friedrich Wilhelm III to house an impressive menagerie on a landscaped island. Wealthy colonial settlers went to great lengths to import exotic animals for sport and other recreational pleasures, but no mid-nineteenth-century private menagerie grew into an Australian public zoological garden. Nor were there any established menageries associated with scientific institutions like those of the London Zoological Society or the Paris Museum of Natural History.

During the 1850s, botanical gardens were established in most of the Australian colonies. Because plants were such an important source of food, fiber, and pharmaceuticals—thereby a source of national wealth—imperial governments were anxious to exploit the diversity of climates in their colonies to establish colonial gardens for the testing of new plants.[1] As part of the British colonial garden network, Australia's botanical gardens were perceived as essential elements in the introduction of new plants into colonial economic production and the selection of colonial agricultural industries that satisfied British as well as colonial needs. Both economic botany and taxonomic botany were well served by these gardens.

While governments were keen to establish gardens for the acclimatization of plants, it was left to private individuals to orchestrate the establishment of organizations for the acclimatization of animals. In some Australian colonies, botanical gardens provided temporary homes for early menageries. Their subsequent permanent homes would metamorphose into zoological gardens that still exist in some Australian capital cities.

The Australian epicenter for acclimatization enthusiasm and activities was Melbourne, the capital of the British colony of Victoria. A series of attempts in the 1850s to introduce game and grazing animals there culminated in the establishment in 1861 of the Acclimatisation Society of Victoria, whose menagerie was initially housed in Melbourne's botanical gardens.[2] The society was responsible for the establishment of Australia's oldest public zoological garden and provided the intellectual momentum for the emergence of sister societies in neighboring colonies. This acclimatization trail led eventually to the establishment of the Royal Melbourne Zoological Gardens.

Mid-Nineteenth-Century Victoria

Nineteenth-century Victoria and other Australian colonies were perceived by their European residents to be sadly deficient in useful and attractive creatures. Nostalgic and material desires required the introduction of new species. The decline in Victoria's goldfields made the prospective rewards from introducing new species particularly attractive.

As a result of the waxing and waning of its goldfields during the 1850s, Victoria changed from a little-known pastoralists' paradise to a bustling, land-hungry colony with a swollen and relatively skilled population to satisfy.[3] Various legislative attempts were made to wrest land from squatters and "unlock lands" for more intensive farming, and serious consideration was given to the stimulation of new rural industries. New ways of exploiting "wastelands" not yet grazed or tilled had to be found, and any plant or animal whose produce could be profitable was eagerly sought.

Individuals with a diversity of skills and interests and a scattering of institutions, including the parliament, the Melbourne Chamber of Commerce, scientific societies, and local newspapers, became involved in the discussion and implementation of animal and plant acclimatization—all in the interest of the colony's agricultural progress. The main orchestrater of Australian acclimatization activities was Edward Wilson, the English-born joint owner and former editor of the Melbourne newspaper, the *Argus*. Wilson was impressed with the philosophy underlying a young French society and was instrumental in establishing acclimatization societies in England and Victoria.

Edward Wilson in Europe

The world's first acclimatization society, the Société Zoologique d'Ac-climatation, was founded in Paris in 1854.[4] This new society concentrated its efforts primarily on the acclimatization of animals. The founding president was Isidore Geoffroy Saint-Hilaire, professor of zoology at the Paris Museum of Natural History and director of its menagerie.[5] In his inaugural address, Isidore Geoffroy Saint-Hilaire expounded the universal and egalitarian ideals of acclimatization—ideals that impressed Wilson. Geoffroy Saint-Hilaire pointed out that the unique association they were about to form

> was to be composed of agriculturists, naturalists, landowners, all the scientific men, not only of France, but of every civilized country, all of whom would aid in a work which required the help of everybody, because it was for the good of everybody. The prospect was nothing less than to people [their] fields, . . . forests, and . . . rivers with new guests; to increase and vary [their] alimentary resources, and to create other economical or additional products. In the vegetable kingdom much had already been done; but in the animal, almost nothing.[6]

After retiring from his position as editor of the *Argus,* Wilson returned to England to seek relief and remedy for his failing sight. In the late 1850s, he provided one of the earlier and more strident voices calling publicly for British action in the field of acclimatization. He was convinced of the value and impor-tance of the introduction of new plants and animals into British colonies and used the London *Times* to make his point. Wilson was no mere armchair acclimatizer: he had already liberated nightingales in Melbourne's Botanical Gardens and was aware of efforts aimed at introducing the alpaca into Victoria. A mixed flock of llamas and alpacas which was for sale outside London in 1858 provided him with an opportunity to effect that introduction. In July 1858 he sought financial support from the British public.[7]

With missionary zeal, Wilson widened his campaign from alpacas to the diverse zoological world. He wrote with conviction of utilitarian and nostalgic justifications for the introduction of plants and animals into British colonies. Using rhetoric similar to that used by Isidore Geoffroy Saint-Hilaire, he argued for the need for people to help increase nature's bounty in Australia. Wilson complained that although Australia's indigenous animals were interesting, they were unfortunately practically useless, providing only "a little sport and [an] occasional meal." With the introduction of sheep, horses, and cattle, humans had already helped correct nature's inadequate provision for their comfortable existence in Australia.[8]

Wilson further believed that residents in the Australian colonies had an equal right to the pleasures enjoyed by their fellow Englishmen back home: the beautiful music of English birds and the thrill of hunting the same game that was available in England. Wilson advised readers of the *Times* willing to donate birds to Australia to do so via the Ornithological Society's island in St. James Park, where they would be cared for while awaiting shipment to Melbourne.[9]

Wilson was successful in attracting British public support to procure all sorts of creatures and transport them to Victoria, but he was less successful with the British government. In endeavoring to convince the colonial secretary, the duke of Newcastle, of the importance of such introductions, Wilson asked him whether he considered "it a scandal upon [the] empire, with possessions in every quarter of the globe, in every latitude, and every climate, to be doing *nothing* as a nation to distribute the good things of each country throughout the others."[10]

Although the duke agreed with Wilson in general, he subsequently declared that a vote of public money for the newly formed English acclimatization society was inexpedient. Wilson condemned the attitude of the British government and contrasted it with the enlightened attitude of the Victorian government, which was prepared to support financially an attempt to introduce salmon into antipodean waters. Since the British government was quite happy to leave it all to private enterprise, Wilson implored wealthy British landowners, whose interests extended beyond the boundaries of their counties, to support such worthy activities in the colonies of South Africa, Victoria, and New Zealand.[11]

Despite the efforts of Edward Wilson and others, British government support—funds or land—was never obtained. After a mere half decade the English acclimatization society's fortunes were waning. In 1866 it was amalgamated with the Ornithological Society.[12]

In Search of the Alpaca

Meanwhile, in antipodean British colonies, dedicated individuals sought the introduction of various game and pastoral animals. While the Melbourne Chamber of Commerce considered the best means of promoting agriculture and settling the wastelands of the colony,[13] articles and letters appeared in the Melbourne *Argus* advocating the introduction of all sorts of new plants and animals into Victoria. Most highly publicized was the alpaca. A hardy, relatively disease-free producer of fine wool, the alpaca had been a contender for introduction into Victoria and its sister colony, New South Wales, since the early 1850s. However, because South American governments prohibited the export of the alpaca, it was extremely difficult to procure. Since it could not legally be taken out through any Peruvian port, the alpaca could only be smuggled out over the awesome Cordilleras to the Argentine Confederation and then into Chile for embarkation.[14]

Thomas Embling, a medical practitioner and politician, shared Wilson's belief in the benefits to be derived by Victoria from the introduction of new plants and animals. While Edward Wilson exhorted the British to fulfill their duty to the Australian colonies by effecting introductions of useful creatures, Embling worked tirelessly toward the same end in Victoria, both inside and outside the parliament. Having already introduced foreign waterbirds to the large lagoon in Melbourne's Botanical Gardens,[15] he was now interested in the alpaca. Added to the difficulty of obtaining the alpaca was the difficulty of obtaining government support. Although the Melbourne *Argus* wrote glowingly of the need to introduce the alpaca into the colony, the government was less than impressed. The 1856 report of a government committee, chaired by Embling, which promoted the importation of alpacas was, in effect, laughed out of the parliament.[16] Even an appeal to Queen Victoria through acting governor Edward Macarthur had no success. Victoria's governor, Sir Henry Barkly, conveyed the queen's disappointment that, since no alpacas remained in her possession, she was unable to help. Titus Salt, the Bradford (England) manufacturer of alpaca wool, however, did have some live animals and had recently shipped four alpacas to his nephew in South Australia. If those animals did well in Victoria's sister colony, then he would "have no objections to sending out a few more to [England's] enterprising colonies."[17]

Meanwhile, Embling persisted. He considered forming "a committee of gentlemen to inquire into the adaptability of animals to [Victoria], with the view of establishing an institute something after the plan of the London Zoological Society."[18] However, the Philosophical Institute of Victoria had just set up a similar committee to investigate the "utility and practicability of introducing the Camel and other useful animals into the Australian Colonies."[19] Rather than duplicate the worthy efforts of the institute, Embling dropped his plans and congratulated the institute.[20]

In case the message had missed rural residents, Embling wrote a long letter to the Port Phillip Farmers' Society extolling the value of the alpaca to the Victorian farmer and also mentioned the less-publicized Rocky Mountain and California sheep and the Great Asiatic argali (wild sheep).[21] Unfortunately, neither the Philosophical Institute's committee nor a company of influential gentlemen with South American connections, which was formed to introduce the alpaca,[22] yielded anything other than transient hope and publicity for animal introductions.

Back in the parliament, Thomas Embling again sought government support. This time the focus was widened to consider "stock." The alpaca remained in his mind but not in the wording of his motion of December 1856:

1. That the introduction of new and valuable stock is essential to the efficient development of the capabilities of this great country.

2. That a Committee . . . be appointed, empowered to take evidence, and to inquire into and report upon the best method of effecting this object, and of the animals most suitable to be beneficially introduced.[23]

Not surprisingly, Embling chaired the committee. The proceedings and report of the livestock committee indicate the dominance of Embling and his ideas. Listed as livestock that would materially benefit the colony, "without invading the pasture grounds of the domestic breeds or even trenching on the liberty of a single sheep," were the quagga, pheasant, partridge, curassow, various deer, antelope, gazelle, eland, gnu, oryx, koodoo, and buffalo.[24]

The livestock report went further than the alpaca report in its demands on the government. In so doing it presaged those of the Acclimatisation Society of Victoria. It suggested that British and colonial governments cooperate in the transfer of useful animals by reciprocating "acts of mutual goodwill." The report recommended that an annual supplementary grant of not less than three thousand pounds be provided for premiums to be awarded by a government-appointed committee to those who would introduce valuable stock. Land as well as money was required: each of the main ports needed a small paddock where newly arrived stock could rest after their long sea voyages. The committee also recommended the provision of a considerable area of land near Melbourne "where experiments on new stock could be tried, their climatizing be accomplished, the stock itself increased, and thereby its permanence rendered more certain previous to its dispersion among the colonists."[25]

In June 1857 Embling twice moved unsuccessfully that the livestock report be adopted. The report received a mixture of ridicule and support and suffered the fate of the alpaca report the previous year. On the chief secretary's assurance that the government would "turn its attention to the adoption of some systematic plan for the promotion of the interests proposed," Embling withdrew the report.[26] Had the government followed the report's recommendations, there would have been little need for further efforts by those such as Thomas Embling, Edward Wilson, and an acclimatization society in Victoria.

Meanwhile, outside the parliament, Edward Wilson used the Philosophical Institute of Victoria as a forum for the discussion and implementation of animal introductions. In 1857 he presented two papers to the institute suggesting that new species were required from outside the colony. He argued that "experimental farms and gardens should be established, in which every plant, as well as every animal that could possibly be found suitable to the colony should be fairly tested."[27] Wilson's paper on the introduction of British songbirds prompted the Philosophical Institute to fund a new committee to cooperate with him in further introductions of birds into Victoria.[28] To house new avian arrivals, the Victorian government agreed to erect a large aviary enclosing several trees in the Botanical Gardens. Wilson was pleased to report that the aviary had soon become "one of the principal attractions" of the gardens. "[A]gainst the wires of this aviary our colonial children flatten their little noses, with a delight . . . as they make their first acquaintance with the lark, linnet, thrush, blackbird, goldfinch, and nightingale, known to them as household words among their parents, but till now never known by sight."[29]

Later, in England, Wilson achieved what Embling had also long desired—the landing on Victorian shores of the highly desirable alpaca. While alpaca smuggler Charles Ledger persevered with his scheme in South America, Wilson continued his well-intentioned harassment of the British. Both were successful. In November 1858 Ledger landed a mixed flock of 276 alpacas, llamas, and alpaca-llama crossbreeds at Sydney, New South Wales.[30] The animals shared the pastures of Macarthur's well-acclimatized merino sheep, whose ancestors had been smuggled out of Spain for the English king George III.

In England, Wilson managed to inspire sufficient subscribers to purchase part of another mixed alpaca-llama flock that had been brought to London via Panama and New York. This gift to the colony arrived in Melbourne in February 1859.[31] In Victoria, Embling made sure that the parliament and the public were aware of this remarkable achievement.[32]

In 1859 Edward Wilson continued to organize the transmigration of all sorts of creatures to Victoria. He dispatched pheasants, thrushes, and skylarks from London to Melbourne and casks to Mauritius for a consignment of the "guaramie [sic], the very best pond fish in the world," which was destined for a special pond in Melbourne's Botanical Gardens. Wilson was also making preparations for the transport of salmon ova to Tasmania. He was busier now than when he had been editor of the *Argus;* it was all becoming too much. He unsuccessfully sought help in collecting information about introduced animals from the Philosophical Institute.[33]

Agricultural progress still needed more injections of new plants and animals. In the absence of an appropriate society in Victoria, a pastoralist, William Lockhart Morton, wrote to the governor, Sir Henry Barkly, suggesting that he request the Foreign Office in London to coordinate the collection of plants and animals into the colony. In his reply, the governor suggested the cooperation of the Philosophical Institute, newly renamed the Royal Society of Victoria. In October 1860 Morton reported this to the Royal Society with the suggestion that the society assume the functions of the French and English acclimatization societies.[34] Morton's ideas on ways of encouraging and coordinating plant and animal introductions were later implemented, not by the Royal Society, but by an acclimatization society whose existence made the Royal Society's committee redundant.

A Zoological Society

In 1857, influenced by the efforts of Wilson and Embling and by ideas aired at Philosophical Institute meetings and in the columns of the *Argus,* two potentially fruitful developments were initiated—an experimental farm and a zoological society. The experimental farm was established largely at the instigation of the Port Phillip Farmers' Society to determine a system of husbandry suited to the soil and climate of the colony. Its aims included the investigation of new plants and animals to determine the most suitable varieties and breeds. Unfortunately, this short-lived venture was not successful.[35]

In the short term, the zoological society was also not successful. In the long term, however, one great success did eventually emerge—the Royal Melbourne Zoological Gardens. In October 1857 a meeting was held in Melbourne to form an ornithological society primarily to breed and exhibit fancy poultry and cage birds. At that meeting, the scope of the society was widened. Dr. Thomas Black, a prominent Melbourne physician, moved "that a society be at once formed under the name 'Zoological Society of Victoria,' whose objectives would include the introduction and acclimatisation of exotic birds and animals."[36]

Capitalizing on the current interest in animal introduction, members of the new society moved rapidly. By early November they had met with the governor and colonial secretary and decided on the objects of the society:

> First: The introduction and improvement of domestic birds and animals, for which exhibitions shall be held periodically within Melbourne, and prizes awarded.
> Secondly: For the importation, care, and domestication of mammalia, fishes, birds, and reptiles of this and other countries, more particularly those of rare and uncommon species.
> Thirdly: The encouragement of singing birds, and the endeavor to propagate them in the country.
> Fourthly: The obtaining [of] a grant of land from the Government for the purposes of the Society.
> Fifthly: It is contemplated that a portion of the gardens be set apart and arranged for the care and protection of birds, mammalia &c., which private individuals may import into the colony, in order that by particular care and attention they may become acclimatised, at such terms as may be agreeable to the discretion of the Society.[37]

The government agreed to provide three thousand pounds for the establishment of a zoological garden.[38] The Zoological Society was also given an occupation grant of thirty acres of land by the Yarra River across from the Botanical Gardens.[39] The Philosophical Institute was also cooperative: it agreed to hand over to the Zoological Society both the birds en route from England and their aviary, which was being constructed in the Botanical Gardens.[40]

A general committee of the Zoological Society of Victoria was formed in January 1858. Its members included respected gentlemen from diverse niches of Melbourne society including legal and medical men. A judge was the president. Edward Wilson and Dr. Thomas Black were members of the council.[41]

In 1858 Melbourne's Botanical Gardens gained an aviary and a menagerie. The menagerie of the Zoological Society was temporarily located in the Botanical Gardens while the society prepared its own gardens. The donated animals—including "emus, a Sumatra, and twelve English fallow deer, some monkeys, coalos [koalas] or native bears, wallabys, Cape Barren geese, native companions, black swans, eagles, gulls . . . &c"—were in the care of Dr. Ferdinand Mueller, Victoria's government botanist and director of the Botanical Gardens.[42] In May 1858 negotiations were conducted with the proprietor of a traveling menagerie over his appointment as keeper of the society's zoological gardens and the acquisition of his animals.[43]

A confluence of unfortunate circumstances led to the young Zoological Society's relinquishing control of its menagerie. The land granted to the society was found to be unsuitable for zoological gardens. It was too barren and swampy. Moreover, following a change in government, the anticipated three thousand pounds was not forthcoming.[44] The society's subscriptions could not cover the cost of garden construction and the maintenance of its menagerie. Government intervention was desperately required.

The government agreed to house the Zoological Society's collection permanently on the higher ground of the Botanical Gardens under the control of a board of management. While the menagerie was housed in the public Botanical Gardens, no entrance fee could be charged. So the government agreed to carry out the objectives of the Zoological Society, including the maintenance of its zoological collection, entirely at public expense.[45] In August 1858 government nominees and members of the Zoological Society were appointed to a board of management committee to oversee the menagerie in the Botanical Gardens. Embling, Black, and Mueller were members.[46] Thus Melbourne's first public Zoological Gardens were part of its Botanical Gardens, with Ferdinand Mueller as the director of both.

Unlike the situation in London, where the Zoological Society established and maintained its Zoological Gardens, in Victoria, the embryonic Zoological Society relinquished control of its zoological collection and faded into inactivity. The Zoological Gardens survived, but, unlike London's Zoological Gardens, animal display and dispersal were the main aims. The animals were neither the subject of scientific study nor the source of funds (derived from public exhibition).

The press was used to solicit help from the public. The *Government Gazette* carried the following notice:

USEFUL AND RARE ANIMALS
The Committee of the Zoological Gardens, Melbourne, will feel thankful to any Captains of Ships, or gentlemen coming to Victoria, should opportunity present itself, if they will endeavor to bring with them useful and rare animals.
The Committee will always be happy to recommend the liquidation of any reasonable expenses incurred in the importation of stock, and presented to their care on behalf of the Government of Victoria.
    By order of the Committee,
    Ferd. Mueller, M.D. Director.[47]

By the late 1850s, the zoological collection had become an interesting attraction in the Botanical Gardens. Originally destined for the experimental farm, Wilson's alpaca-llama flock from London joined the menagerie, as did pure alpacas from Ledger's flock.[48] A wooden house sheltered the llama-alpacas and angora goats, and an enlarged aviary housed a symphony of songbirds. The collection comprised

> Angora goats, fat-tail sheep, llama-alpacas, 13 fallow deer, with 3 fawns, a Sumatra deer, Ceylon elk, several kangaroos, and emus, koalas, an ichneumon, monkeys of various species, a considerable variety of singing birds, of which the canaries, goldfinches, and linnets . . . reared broods, while the thrushes [were] nesting; Californian quail, which also increased; native companions, 16 black and 6 white swans, English and silver pheasants, Murray pheasants, Australian eagles, hawks, and several other smaller animals.[49]

Thus, by the turn of the decade, Melbourne did have government-funded zoological gardens. Many animals were breeding in captivity, and some birds had been liberated. Scientific, public, and government support for animal introduction and acclimatization was running high. The Royal Society of Victoria (formerly the Philosophical Institute) continued to provide a forum for the discussion of animal and plant introductions. Merely by strolling through Melbourne's Botanical Gardens the public could marvel at the fruits of these introductions, while both Melbourne and London newspapers applauded them.

This was the scene to which Edward Wilson, still keen to improve the colony by introducing all sorts of birds and beasts, returned from England late in 1860. Wilson must have been pleased to see the stage so well set for the birth of an acclimatization society in Victoria; the defunct Zoological Society appeared the perfect primordium for it.

The Acclimatisation Society of Victoria

Edward Wilson was a man with a mission. By early 1861 he had attracted most of the members of the Zoological Society and many others to join a new society, the Acclimatisation Society of Victoria. The new society would continue the work started by the Zoological Society, but its aims would be much loftier. No longer would the main aim of introducing new animals be their exhibition. The Acclimatisation Society would introduce exotic plants and animals primarily for their dispersal to suitable parts of the colony. Fish, songbirds, and game birds and animals would be "loosed" into the wild; crop plants and grazing animals such as the angora goat and the alpaca would find homes on rural properties.[50] The year of Isidore Geoffroy Saint-Hilaire's death in France saw the birth of a society for the realization of his ideas and aspirations in Victoria.

On February 25, 1861, the Acclimatisation Society of Victoria (ASV) was formally established at a public meeting presided over by the governor of Victoria, His Excellency Sir Henry Barkly. It was resolved that members of the Zoological Gardens Management Committee and Edward Wilson would act as a provisional committee of the ASV and that when properly organized, the new society should approach the government to learn "to what extent it [might] calculate upon Government assistance and recognition."[51]

On condition that the Zoological Gardens Management Committee and its activities be absorbed by the Acclimatisation Society, the government agreed to provide both land and money.[52] In its first year the ASV planned and prepared its zoological gardens and transferred to them most of the menagerie from the Botanical Gardens. It sought and tended an increasing diversity of animals and exchanged animals with London, Paris, St. Petersburg, Mauritius, Calcutta, Colombo, and the Cape of Good Hope.[53]

Melbourne's Zoological Gardens

The land provided by the government was Royal Park, which was then outside the northern limits of Melbourne. Most of that park has survived as parkland and is now Melbourne's largest inner-suburban open space. In March 1862 the Lands and Survey Office gave official notification of the government's intention to reserve permanently for zoological purposes "the reserve known as Royal Park (less that portion thereof occupied by the Experimental Farm)—five hundred and fifty acres."[54] The trustees' first act was to select and enclose an area of fifty acres in the middle of Royal Park for the ASV's depot. One hundred and thirty years later, this area has survived as the site of today's Royal Melbourne Zoological Gardens.

With the fencing of its new site, the ASV could now contemplate activities precluded by the confines of cages in the Botanical Gardens—even secluded breeding areas could be planned. The park outside the fenced enclosure provided extensive daytime pastures for the society's grazing animals and wilderness for the liberation of birds and animals. The first residents of the ASV depot came mainly from the menagerie in the Botanical Gardens.[55] Animal donations continued to be sought from the public via notices in the press. Information and animals were also obtained with official British help. Lords of the Admiralty promised that His Majesty's ships would "be rendered available for the purpose of conveying animals, provided no expense be thrown upon the department."[56] A questionnaire was circulated via the Foreign and Colonial Office to British consuls and governors to elicit information about plants and animals worthy of acclimatization.[57]

Acclimatization was a worthy activity and thanks to the provision of government land and money and to private subscriptions, entry to the ASV depot was free. But the public wanted to see exciting and expensive animals. Alpine alpacas were not enough. Ironically, the fate of the alpacas reflected the fate of acclimatization. As their popularity and health waned during the 1860s, so did that of acclimatization. By 1864 climate and disease had reduced the health and numbers of the llama-alpaca flock in Royal Park. Government grants declined from the initial grant of three thousand pounds; in 1869 there was no grant at all, and staff at Royal Park were paid from a loan raised on a personal guarantee of an ASV council member.

Although interest in acclimatization waned, public interest in zoos did not.

The 1875 plan of the gardens of the Zoological and Acclimatisation Society of Victoria in Royal Park. (Reproduced from the report of the annual meeting of the society, February 26, 1875, published in the report of the board on the Royal Melbourne Zoological Gardens, June 30, 1977)

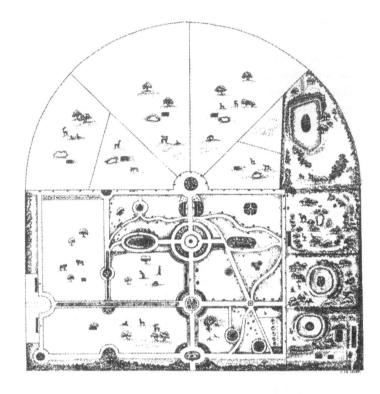

The 1977 plan of the Royal Melbourne Zoological Gardens. The basic outline of the 1875 plan of the gardens of the Zoological and Acclimatisation Society is still recognizable. (From the report of the board on the Royal Melbourne Zoological Gardens, June 30, 1977)

Once the society recognized its zookeeping role and in 1872 became the Zoological and Acclimatisation Society, its financial situation and zoological gardens improved. Inserted into the basically unchanged objectives of the society was "the collection and maintenance of zoological specimens for exhibition."[58] Now the Royal Park depot could be converted into publicly acceptable zoological gardens.

In the early 1860s the concept of acclimatization radiated from the ASV in Melbourne to stimulate branch societies in Victorian country towns and similar societies in neighboring colonies. Some Victorian towns desired zoological gardens rather than acclimatization societies. Commitment to acclimatization was not enduring in other Australian colonies, whose ephemeral acclimatization societies faded before establishing lasting zoological collections.

The ASV had arisen on the crest of a wave of enthusiasm for animal and plant acclimatization. Melbourne's Zoological Gardens were spawned on that enthusiasm and prompted the necessary government provision of land and money. The original collection at the Royal Park Zoological Gardens did not reflect zoological diversity, nor did it excite the public. Rather, it reflected the hopeful enthusiasm for acclimatization and comprised potentially useful animals such as camels, alpacas, and angora and cashmere goats awaiting pastoral homes and deer, hares, quails, ducks, trout, and carp awaiting liberation in the wild.

Anglers and deer hunters may thank the ASV for their recreational pursuits. It is ironic, however, that the most widely known surviving offspring of the ASV is an institution that it never intended—a real zoo: the Royal Melbourne Zoological Gardens.

RAM BRAMHA SANYAL
AND THE ESTABLISHMENT
OF THE CALCUTTA
ZOOLOGICAL GARDENS

*N*ineteenth-century India experienced a florescence in the study of natural science. In the first quarter of the century, progress in this field was mainly due to the work of several dedicated European scientists. Indians, however, quickly started taking an interest in botany and zoology.[1] The first systematic zoological study in India was initiated soon after the establishment of the Calcutta Zoological Gardens in 1876. Participation and involvement by Indians resulted in the establishment of a network of scientific institutions from Bombay to Delhi to Calcutta by the late 1800s. At the turn of the century, scientific societies had become deeply rooted in the new, emerging scientific atmosphere. The Indians who had worked with Europeans had embraced the spirit of free inquiry, and they had acquired experimental skills that enabled them to have an impact on the study of natural history in the subcontinent.

The Establishment of the Calcutta Zoological Gardens

Many previous plans for a zoological garden in Calcutta, including one that was proposed by Sir Joseph Fayrer (1867), did not materialize. Carl Louis Schwendler of the Indian Telegraph Department brought the subject to the attention of Calcutta's Asiatic Society in March 1873.[2] He proposed that the necessary capital be raised by subscription but that the government of Bengal should grant the site and meet the annual expenditure. Incidentally, Schwendler had a fine collection of living animals of his own. He showed it to the then viceroy of India, Lord Thomas George Northbrook,[3] and the viceroy was impressed by how easy it was to maintain such an establishment in a climate so well suited to animal and vegetable life as that in Calcutta. He was of the opinion that instead of large and expensive houses in such a climate, simple sheds were sufficient protection for the animals; the luxuriant vegetation would add to their shelter. Schwendler believed that if space was provided and sufficient fencing obtained, then the animals' new homes would be similar to their original habitat. Schwendler convinced the viceroy, and in turn the viceroy convinced the lieutenant governor, Sir Richard Temple, that Schwendler's approach to keeping animals was both practical and possible.

Sir Richard Temple was a man of action, and in 1875 he granted a large tract of land, 163 bighas (approximately 218,141 square meters), on the road leading from the Zeerut Bridge to the governor's residence at Belvedere for this purpose.[4] An honorary managing committee was appointed in December 1875: Lord Ulick Browne was president; Carl Louis Schwendler and Dr. George King of the Calcutta Botanical Gardens were members; and C. Buckland was honorary secretary. Buckland was at that time private secretary to the lieutenant governor.

In the initial stage, the Zoological Gardens required at least thirty thousand rupees for its layout and development.[5] This amount was raised quickly through donations by Indian princes, nobility, and gentry. Schwendler provided twenty-two of his own mammals to start the animal collection. Soon after, eight other mammals were presented by Indians to the new Zoological Gardens.[6] By late 1875 the gardens had a number of different species. On December 27, 1875, during a visit to India, King Edward VII inaugurated this new institution.

At its founding, the objectives of the Calcutta Zoological Gardens were (1) to provide recreation, instruction, and amusement for all classes of the community; (2) to facilitate scientific observations of the habits of animals, especially those peculiar to tropical climates; (3) to encourage the acclimatization, domestication, and breeding of animals and to improve indigenous breeds of cattle and farm stock; and (4) to promote the science of zoology by the interchange, import, and export of animals.[7]

From the start, the management of the Zoological Gardens was in the hands of Carl Louis Schwendler, electrician to the government of India; Dr. George King, superintendent of the Calcutta Botanical Gardens; and Dr. John Anderson, superintendent of the Indian Museum, Calcutta. This trio constituted the members of the newly formed Zoo Management Committee. King and Anderson both had medical backgrounds. Although science professionals gave rise to the gardens, zoo organization and animal husbandry were based on practical experience. In a few years, the Calcutta Zoological Gardens acquired a good reputation in England, Australia, and several European countries—mainly because of Schwendler and his long experience in managing exotic animals. He was ably supported by King and Anderson and also by a hitherto unknown Indian, Ram Bramha Sanyal.

After the phase of initial establishment, the committee thought seriously of employing a director. A long search for a proper scientific person from England or from India, however, proved fruitless. Ultimately, an employee of the Peninsular and Oriental Steam Navigation Company, J. C. Parker, who had previous experience with animals and birds, was appointed in 1876.[8] His services lasted only for six months because of the poor financial resources of the committee and the government of Bengal's refusal to fund this appointment.[9]

At this time, a young Indian from Bengal, Ram Bramha Sanyal, who ultimately became the most influential figure in the gardens' early period, was recruited in a very junior capacity to assist King. A few other Indians were also employed in posts such as gardeners, watermen, watchmen, and gate clerks. Although these Indians all came from different walks of life, they worked as a team for the newly laid out gardens. The living conditions of the small number of keepers were simple, and they were accommodated within the gardens.

When the Zoological Gardens were established, the government of Bengal promised to pay twenty thousand rupees a year as a grant. For reasons unknown, however, it stopped the funds between July 1876 and July 1877.[10] Since the gardens' financial position for the first few years was shaky, the committee had to initiate ways and means to strengthen its economic position. Thus the zoo functions became geared not only toward education, research, and conservation but also for public entertainment. For example, in August 1883 two natives from the Andaman and Nicobar Islands were brought to Calcutta to serve as a living exhibit (part of the Calcutta International Exhibition). They were paid to stay in the Zoological Gardens during the daytime under the shade of a large tree. People came in large numbers to see them, willingly paying the required gate fee. For some time they were a major attraction.[11] Moreover, revenue earned in the form of gate receipts and fishing and boating fees was a big factor in strengthening the financial position of the zoo.

Departing from Schwendler's original "simple sheds" concept, the animal houses and enclosures, erected over a course of years on the widespread grounds, were similar to the buildings at the London Zoo. At least one of the committee members was invariably visiting London once a year for his annual leave, thereby gaining intimate knowledge of the animal houses and cages at that city's zoo. This seems to have influenced the committee in its construction plans for the new gardens.

The Calcutta Zoological Gardens' eight employees gradually increased as the workload increased. The Zoo Management Committee, which supervised the employees, was also enlarged. In addition to Schwendler, King, and Anderson, the committee now included S. A. Stewart (executive engineer), Major Charles Mant, Major R. C. Strandale, Dr. D. B. Smith, and H. M. Tobin. Of them, Major Charles Mant, of the Royal Engineers, was skilled in providing designs for various cages and buildings. The weekly meetings had a unique way of solving critical issues and day-to-day problems. Schwendler, Anderson, and Smith were responsible for the animals' comfort, health, and so on. King was basically responsible for the landscape gardening. At first, Strandale was given the charge of arranging food for the animals; however, later on the committee decided to entrust the work to Sanyal.

In the late 1870s and 1880s, the Calcutta Zoological Gardens' animal list included African buffalo, Zanzibar rams, domestic sheep, four horned sheep, hybrid Kashmiri goats, Indian antelope, Indian gazelles, Sambhar deer, spotted deer, and hog deer. Carl Schwendler alone (up to January 11, 1880) presented 31 mammals and 57 birds. Within a year, the collection began to swell owing to animal donations, exchanges, and purchases; this was coupled by the shift of Barrackpore Park Collection to the gardens by the middle of 1876.[12] By April 1877 the gardens had 756 mammals, birds, and reptiles.[13] Only a year later, the number of mammals (tigers, leopards, deer, bears, and so on) had so increased that the committee decided to sell some of them at suitable prices.[14] Sanyal, with the help of committee members, initiated a collection policy to control the haphazard manner in which the animals had been acquired. Moreover, Sanyal believed that the objective of the zoo could never be simply to put animals on exhibit; showing naturalistic behavior was the real goal. He helped to establish successful breeding programs for the mongoose, lemur, short-spined porcupine, agouti, tiger, and leopard.[15] The zoo's emphasis had switched from showing off the maximum number of first importations to sustaining breeding groups.

Ram Bramha Sanyal

The history of the zoo in its first thirty years centers around Ram Bramha Sanyal. In the absence of a European director with a scientific background, the committee had to depend solely on an Indian, Sanyal, who proved equal to the task.

Sanyal was admitted to the Calcutta Medical College probably in 1870. Unfortunately, his medical studies were cut short after three years. His eyes started giving him trouble,[16] and he was warned by his professors that if he persisted in his studies, he might go blind. Sanyal gave up his studies. The Medical College, however, was where Sanyal came into contact with Dr. George King, superintendent of Botanical Gardens at Howrah.

As a member of the Management Committee, King was given the responsibility of landscaping the new Zoological Gardens. The actual area acquired for the gardens was slightly more than 156 bighas (208,773.24 square meters). Naturally, for this extensive area, he needed help. He hired a number of Indian laborers and workers and chose Ram Bramha Sanyal to supervise them. The young lad entered as a casual worker on January 24, 1876, probably on a daily wage basis.

On May 1, 1876, the gardens were opened to the public. There was a wooden post fence around the boundary. In the first week, there were only 82 visitors. With the advent of the cooler season, however, the number of visitors increased; in December 1876, 10,874 people came to the gardens.[17] The main workload was on Sanyal; his job was to monitor everything that took place within the gardens.

By September 1876 the committee promoted Sanyal to head babu (head assistant), increasing his pay to forty rupees per month (probably by recommendation of King, for whom Sanyal worked for several months). In August 1876 Sanyal assumed more responsibility when the animal collection of Barrackpore

Park was shifted to the gardens. By direction of the committee, the management, food, and health of the 756-specimen collection were to be supervised by Sanyal.

From January 1877 the Zoo Management Committee felt that, as the collection was fairly large, regular observations had to be kept for future reference. It decided that the observations would be recorded in a "Daily Register." The entries to the Daily Register were entrusted to Sanyal. The work was of a technical nature and was certainly not expected to be done by an ordinary worker. Sanyal rose to the occasion. In his Daily Register, Sanyal included animal activity, noting even urination and defecation habits. (Interestingly, the Daily Register of January 1877 does not include mention of the escape of two tigers, which had to be destroyed the following day.) In a committee meeting of June 21, 1877, members felt that Sanyal's duties were too numerous and that a babu might be employed to assist him, especially in his supervision of the gardening work.

Despite Sanyal's best efforts to satisfy the Management Committee, however, the members, in their July 19, 1877, meeting, were still hoping for a European superintendent. They felt that Sanyal was "unfit to have the job of management of the Gardens." They wanted to bring in someone from England.[18] When the committee's secretary, H. M. Tobin, went to England to try to select a person for the gardens, he could not find anyone suitable. Thus additional duties were assigned to Sanyal.[19]

Since the Zoological Gardens were one of the new recreation attractions in Calcutta, they were visited by all kinds of people, including high government officials. Sir Alfred Croft, director of public instruction for Bengal, was one such visitor to the gardens. He often met and talked to Sanyal during his visits and soon offered Sanyal a job. Sanyal submitted his resignation to the Management Committee on July 25, 1878.[20] The committee, however, believed that Sanyal's departure might create a vacuum and undermine the implementation of the final

Ram Bramha Sanyal at age thirty. Taken in Calcutta more than twenty years earlier, this photograph appeared in a local Bengali magazine in October 1908. Sanyal died on October 13, 1908. (Author's collection)

layout of the gardens. Sanyal was not allowed to join the new service.[21] This had been an attractive offer for Sanyal; it was a government service job, which meant stability in his life. Soon thereafter, C. T. Buckland, president of the committee, took the initiative and declared the Calcutta Zoological Gardens a government institution beginning on October 10, 1878. This move was sanctioned by the lieutenant governor of Bengal and the government of India. The gardens would now become a government institution similar to the Calcutta Botanical Gardens.[22] Thus Sanyal got the stability he desired. He was appointed acting superintendent in April 1879 and was made permanent in the post in April 1880.[23]

Visitor fees collected at the gate were never enough to defray the gardens' expenditures. Finances went from bad to worse. The total annual grant from the government was twenty thousand rupees, but it was insufficient to support the expanding facilities at the zoo. In March 1879, in different committee meetings, the members decided to hold various shows and exhibitions in the gardens by selling tickets.[24] Sanyal had to supervise these productions, making sure that they were popular with visitors but created no ill effects on the animals.

According to the original prospectus of the gardens, one of the principal objectives was to encourage the acclimatization, domestication, and breeding of animals and to improve indigenous breeds of cattle and farm stock. In June 1882 the Committee resolved to start a dairy farm in the adjoining land, Begumbari, an area of 43 bighas (57,546.59 square meters). Approval of the venture was obtained from the governmental authorities.[25] Measures were taken to bring cattle from Australia for the improvement of Indian breeds. The committee sought to procure four young elephants to dispatch to the Melbourne Acclimatisation Society in exchange for the cattle. Unfortunately, the farm did not work out for a variety of reasons, including its marshiness.[26] In this venture, too, Sanyal was to see that the land was properly enclosed with iron fencing, that sheds were made according to the standard measurement, and that the cattle were regularly inspected.

As a reward for his years of hard and honest labor toward the smooth running of the gardens, the Management Committee decided in April 1902 to request the government of Bengal to appoint Sanyal as a member of the committee. Sanyal joined the committee on June 21, 1902. During his tenure, he suggested many ways to improve the gardens. Until his last day, he worked with a deep insight for the betterment of the zoo's animals and his co-workers.

Recognition of Ram Bramha Sanyal's Work

A new lieutenant governor, Sir Stuart Colvin Baylay, arrived in Calcutta in 1887. In the Management Committee's Annual Report of 1888–89, Baylay suggested that, on the basis of old records and recollections of the superintendent, a handbook be produced which would help innumerable persons having animals and birds in captivity.[27] Finally, in March 1892, *A Handbook of the Management of Animals in Captivity in Lower Bengal*, written by R. B. Sanyal, was published by the committee. This work was highly appreciated by all in this field.[28] In the next year, as a recognition of Sanyal's work, the London Zoological Society made him a "corresponding member."

In March 1894 Sanyal was invited by the secretary of the Bombay municipality to visit the Zoological Collection at Victoria Gardens (Bombay).[29] Bombay officials wanted advice on their expansion plans. Sanyal submitted a detailed report suggesting a number of improvements for the Victoria Gardens Zoo. He included in his document a sketch map of various proposals including their measurement, probable cost, and maintenance features.[30] At the suggestion of Sanyal, J. M. Doctor, head animal keeper, was assigned by the municipal commissioner of the city of Bombay to Calcutta for one year in order to be trained by Sanyal at the Calcutta Zoological Gardens.[31] Certainly this was an unparalleled recognition of the knowledge and depth of Sanyal's understanding of the management of animals in zoos.

In 1896 Sanyal wrote his second book, *Hours with Nature*.[32] Written in simple language to stimulate youngsters' interest in the natural world, it was meant mainly for school students.

As a true scientist, with brigade surgeon Lieutenant Colonel D. D. Cunningham, a member of the Management Committee, Sanyal conducted experiments in 1895–96 on the action of various reputed antidotes to snake venom.[33] These experiments had great significance. Even when Cunningham retired, Sanyal continued his experiments in the zoo laboratory.[34] With his increasing success in "zoo biology," the committee sent him to Europe in June 1898 to get firsthand knowledge of various zoological gardens. He also joined the Fourth International Congress of Zoology (Cambridge) in August 1898.[35] On his return to India on January 1, 1899, he was awarded the title of "Rai Bahadur" and was soon made an associate member of the Asiatic Society of Bengal, Calcutta.[36] He often presented scientific papers at the society's monthly meetings.

In mid-1899 Sanyal lost his only son and then, within a year, his wife as well. His health was shattered, and he was physically and mentally weak for a long time. His work was affected. Even in this condition, he continued to attend monthly meetings of the Asiatic Society and write articles in Bengali for children. Somehow, Sanyal carried on and continued his effective supervision of the gardens.

In 1902, however, he hinted to the Management Committee that he might retire because of his failing health. In 1904 it became a reality as Sanyal applied for his retirement and pension starting January 1906.[37] Captain Harold Brown, then secretary of the Management Committee, requested the government of Bengal on February 6, 1904, to arrange for the training of another person for the post of superintendent within the next two years.[38]

On the recommendation of Gerald Bomford, surgeon general of the Indian Medical Service, Babu Pasupati Mitra, an additional demonstrator of anatomy at the Medical College in Calcutta, was nominated for the post of superintendent.

In addition to his knowledge of human anatomy, Mitra also had a veterinary degree. Naturally with his qualifications, the committee felt that he was the most suitable person for the post.[39] Mitra was employed as assistant superintendent on April 20, 1905, but with the condition (demanded by the government of Bengal) that his initial appointment be only for six months.[40]

Unfortunately, Mitra was transferred back to his parent department as of November 27, 1905; he was losing the opportunity of promotion in the Medical College service. Sanyal was requested by the committee and by R. W. Carlyle, chief secretary to the government of Bengal, to stay for another two years, until they could get a competent man for the post. Finally, after a long, fruitless search, the committee requested Sanyal's guidance. Sanyal recommended Bijoy Krishna Basu, a veterinary inspector from Assam. Basu was finally placed in the post of assistant superintendent beginning April 1907.[41]

Sanyal got year-to-year extensions of his service. His final extension was up to February 14, 1909.[42] By October 1, 1908, Sanyal was not in good health,[43] and he applied for five weeks of privileged leave beginning October 12. He remained in his Zoological Gardens residence. On October 13 he died; his inseparable link with the Calcutta Zoological Gardens came to an end.[44] He left behind many persons to mourn him and also a large number of animals whom he had loved most. The history of the Calcutta Zoological Gardens in its early part ends with the passing away of Ram Bramha Sanyal.

Sanyal's qualities were varied and numerous; he had a combination of natural ability and practical versatility which resulted from a good knowledge of certain subjects. He was proficient in zoology in general and the treatment of animals in captivity in particular. He was an efficient pathologist, a thorough sanitarian, a capable gardener, and a decent manager with firmness of character which commanded the respect and obedience of his staff. Sanyal was a naturalist with a genuine affection and sympathy for animals; in his handling of them, his first concern was their health and general condition. Through his experience, he found that there were no universal rules for the treatment of wild animals in captivity. Even individuals of the same species had to be managed according to the particular circumstances of each case.[45]

It can be said without hesitation that Sanyal's success was unique, unparalleled in the history of science in India. There is no denying the fact that Sanyal was at the top of the Indian science scene for some thirty years. During this time, only a few selected and aspiring Indian persons could think of making science their profession.

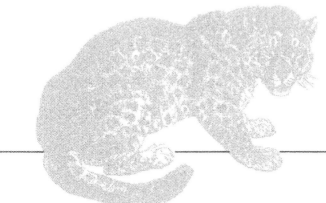

THE AMERICAN SCENE

AMERICAN SHOWMEN
AND EUROPEAN DEALERS
Commerce in Wild Animals
in Nineteenth-Century America

*D*uring the 1800s, both American scientists and showmen sought to
present zoological and aquatic specimens to the public. Their
displays were made possible because of the development of trans-
atlantic trade networks and the growing market for exotic wild animals in
nineteenth-century America. In this chapter, I examine the nineteenth-century
advances in trade and marketing which underlay such animal exhibitions.

In his extensive three-volume history of menageries, Gustave Loisel reported that the Dutch East India Company was the major supplier of animals to
Europe in the seventeenth and eighteenth centuries. In addition to constructing
special pens and stables at an Amsterdam dock, the East India Company established a depot and menagerie at the Cape of Good Hope which existed until
1832. While the East India Company seemed to have had a sideline business of
animal importing, a large number of creatures seem to have arrived in Europe as
the speculative ventures of individual shippers.[1] Thus it was left to some menagerie operators to serve also as brokers to other buyers.

In London, Edward Cross was an important dealer at his famous Exeter
'Change Royal Menagerie. The menagerie was originally established sometime
in the 1770s by Gilbert Pidcock as the headquarters for his show when not
traveling during the summer fair season. Cross acquired the menagerie in 1817.
One of its most famous attractions was the elephant Chunee. Unfortunately,
Chunee was shot on March 2, 1826, because of a growing concern about his
annual fits and the danger of his destroying cages and letting loose other wild
animals.[2]

In 1822 Chunee's keeper, Alfred Cops, took over the famous Tower Menagerie, which was established in the twelfth century and famous for its collection of lions. Cops transformed the languishing collection but was ordered to
remove it from the Tower in 1835.[3] For a time, Cops was a useful contact for
American showmen seeking wild animals in the European market, and his home
in the Tower occasionally provided lodgings for his American visitors. Such

frequent and familiar contact eventually resulted in his daughter's marrying an American showman in 1841.[4] One American buyer in London in 1833 summarized the state of the animal trade there: "[T]he animals offered are mostly common in kind or inferior in quality."[5] Thus enterprising American exhibitors would have to look to several sources for animals to exhibit.

The first elephant brought to America was imported by a ship's captain in 1796, but no records exist of its travel after 1799.[6] A second elephant arrived in Boston from the East Indies during the spring of 1804 and traveled extensively until it was shot in Alfred, Maine, in 1816—a widely reported event.[7] Within two months, owner Hachaliah Baily was arranging for up to eight thousand dollars to replace his loss but this time seeking a male elephant ("preference being given in the selection to one that is white") to be shipped from Calcutta.[8]

The period from the late 1820s until the Panic of 1837 was an especially good era for American menagerie entrepreneurs. In the early 1830s elephants were no longer the single most desirable attraction. In 1831 showmen began to feature a new animal—the rhinoceros.[9] While James R. Howe, co-owner of the New York Menagerie, was in London bargaining with animal dealers who were asking one thousand pounds for a rhinoceros, a new and competitive era had begun among American showmen which involved international financial arrangements with such banking houses as Baring Brothers in London.[10]

By the mid- to late 1830s, at least two American expeditions traveled to Africa in search of yet another novel animal, the giraffe. The first two giraffes were imported to the United States by Rufus Welch in 1831 from Cape Town. Later, groups came via Egypt, including several giraffes captured by Benjamin Brown, who spent thirteen months collecting the animals in the Sudan region of Africa.[11]

In 1834 the New York Menagerie owned by James R. and William Howe toured New England. A rare surviving inventory made at the end of the season provides insight into the menagerie's $46,000 worth of animals and equipment. Columbus, the elephant, was valued at $9,000, and the rhinoceros was worth $10,000. A zebra, Bactrian camel, gnu, two tigers, and a polar bear were each worth from $2,000 to $3,000, but parrots and monkeys were valued at only $15 each. Fifty-two horses, fifteen cages and wagons, a candy wagon, and a wagon for the band and performers completed the show.[12]

The inventory was probably prepared as part of the appraisal for a joint stock company called the Zoological Institute which merged all nine of the menagerie companies in the United States. The total value of all the animals and equipment in the new company was $329,325, and on January 14, 1835, at the Elephant Hotel in Somers, New York, 123 investors subscribed to the stock. Like the several museums of the Peale family, the Zoological Institute was formed "for profit" but also with the belief that "the *knowledge of natural history* [might] be more generally diffused and promoted, and rational curiosity gratified."[13] The institute controlled all thirteen menageries that toured in 1835; a few were combined with circus companies. The grand consolidation of 1835, however, ended with the Panic of 1837. P. T. Barnum, who was born in Connecticut only a few miles from Somers and had occasionally worked for some of the Zoological Institute showmen early in his career, recalled in 1854 that the institute failed because it was conceived by speculators who schemed to sell the stock, pocket some profits, and then let the stockholders look out for themselves.[14]

These early menageries also served a useful function in a young country that placed a high value on knowledge. While the schools were more often concerned with children's moral development than with the development of their minds, publications relating to natural history began to appear in the early nineteenth century.[15] Not only did the menagerie shows expose a broad spectrum of Americans to examples of the animal kingdom, but by 1836 they also widely distributed booklets describing the animals on exhibit. These informative souvenir booklets referred to explorers' accounts and cited the works of eminent natural scientists such as the Frenchman Buffon.[16] A larger cultural world was reaching the countryside through the traveling animal shows.

The Panic of 1837, followed by the depression of the early 1840s, doomed many big wild animal shows. Whether the novelty was wearing off for the public or the cost of importing and providing daily feed and care for the beasts was too burdensome, the number, size, and variety of animal caravans shrank considerably. Major showmen such as Rufus Welch, importer of the first giraffes, virtually abandoned the menagerie business in favor of circus management. The switch was logical from a management viewpoint, since the operations were similar: both shows moved daily and required heavy wagonloads of exhibition paraphernalia. Only one major, large-scale menagerie endured through this period and into the decades after the Civil War. The American proprietors of the Van Amburgh menagerie had toured England and the Continent in the early 1840s but returned to the States in 1846. The show, named for its first star animal trainer (who died in 1865), traveled by horse-drawn wagons for nearly four decades with an outstanding collection of animals which generally received top billing over its occasional circus features—often boasting of itself as a strictly "moral show."

After the Civil War, competition in the American circus business intensified, and rival show managers, perhaps recalling the boom years of the big menagerie shows, expanded their collections of wild animals. Unlike Europe, where the circus and menagerie remained two separate enterprises run by different families, in America the two attractions were now closely associated and the menagerie was an integral part of the show—one ticket admitted the holder to both the menagerie and the circus. One result of the renewed menagerie that was now part of the circus was the emergence of the specialized dealer in wild animals. Showmen, faced with the expanding complexities of their perambulating exhibitions, no longer chose to be involved in the great financial risk of outfitting expeditions. Moreover, a competitive market was evolving which could not depend on the casual shipboard arrival of wild animals.[17]

Meanwhile, fledgling efforts were under way in the decades after the Civil War to open the first zoos in some of America's larger cities. By the late nineteenth century, fewer than a half dozen American zoos had any large animals such as an elephant, since most were little more than deer parks. Circuses had far better collections and, of course, reached a larger and geographically broader audience. In fact, the first elephants (in 1891) and the first lion at Washington's

An 1860 newspaper broadside from Rome, New York, advertising the coming of the famous Van Amburgh animal show, one of only a few traveling menageries to survive the Civil War. The note in fine print at the bottom reads: "The Management would respectfully state that in organizing their Circus Company for the Campaign of 1860, they have spared neither time, labor nor money to make their present combination the MOST BRILLIANT and ATTRACTIVE ever presented for the patronage of the public." This kind of broadside was typical of circus advertising in the United States at that time. (Collection of Jane Mansour)

Van Amburgh and Company's "mammoth" elephant, Hannibal, in August 1860. This pachyderm was promoted as the largest in the world. According to a presentation inscription between two circus men on the reverse of the original card-mounted photograph, this large Asian elephant was imported by George Vaughn of New York City and sold to menagerie operator James Raymond in 1833. At its death in 1865, this elephant was nine and a half feet tall. Believed to be the earliest known photograph of a live elephant, this image was taken on August 18, 1860, in Newport, Vermont, and the photograph was printed by Howes of Brewster, New York. (Author's collection)

National Zoo came from circuses, as did the zoo's only Sumatran rhinoceros (in 1893).[18]

In the United States, the leading firm was headed by two German brothers, Charles and Henry Reiche. About 1844, still in their teens, they immigrated to New York and began to peddle birds. By 1853 twenty thousand canaries had been imported by the Reiches, and in that same year Charles published *The Bird Fancier's Companion, or Natural History of Cage Birds, Their Food, Management, Habits, Treatment, Diseases, Etc.* A third edition appeared in 1867, followed by seven more printings in the next four years. This demand clearly reflected Victorian America's preoccupation with pets. Early prosperity was assured for the Reiches when they shipped three thousand canaries to San Francisco sometime after the gold rush fever began. The yellow birds were rapidly exchanged for yellow metal at twenty-five dollars each, and the price soon jumped to fifty dollars apiece. The Reiches' large bird trade brought them in touch with showmen, resulting in the famous Van Amburgh show's asking them to import other animals. Henry Reiche began to watch the sales of surplus animals by European zoological gardens and to buy such beasts as he was able. Soon, the post–Civil War rivalries of American showmen caused a demand that could not be supplied immediately, and even the Reiches were competing with their own customers.[19]

For example, when the English menagerie of Mr. and Mrs. Alexander Fairgrieve, one of three founded by Mrs. Fairgrieve's uncle, George Wombwell, was auctioned in Edinburgh, Scotland, in 1872, a local newspaper reported that the buyers were "Mr. Jamrach, the most extensive dealer in wild animals in the world; while next to him was a quiet, modest-looking gentleman, who turned out to be the great Jamrach's almost equally great rival, Mr. R[e]ic[h]e." Also attending was William Cross, the Liverpool-based grandson of the famous Exeter 'Change showman and dealer, as well as representatives of zoological gardens in Manchester, Bristol, and Paris. Among the prominent menagerie showmen who were buying was "Mr. Ferguson, the representative of the famous Van Amburgh, who has now three menageries in America." Ferguson's delighted partner, Hyatt Frost, reported in a letter that the twenty-two large cages were "the largest and best collection of animals ever imported to this country by any private enterprise" and included the first black African rhinoceros brought to America and only the second seen in modern Europe.[20]

The longtime New York shop of Charles Reiche and Brother was at 55 Chatham Street near the Bowery. Each side of the store was lined, floor to ceiling, with bird cages. A thirty- by twenty-five-foot backyard housed box cages stacked on one side and crowded with lions and leopards; opposite these cages was another ramshackle pile of boxes with monkeys for sale as pets or to be trained to work with organ grinders. In the rear, extending ten feet farther back, was an enclosure housing three or four gnus, according to one reporter in 1870. About 1860, while Henry stayed in New York, Charles Reiche moved to Alfeld,

Germany, and remained abroad until his death in 1885. Alfeld is near where the brothers grew up, but, more important, it is only a hundred miles due south of the port of Hamburg, a longtime animal trading depot. The Reiches are credited with being among the first to turn the casual animal dealings at the Hamburg port into a more formalized business. The Reiches also maintained agents at principal shipping points such as Egypt and Ceylon, where their agents both recruited native hunters and arranged for shipments to Hamburg. From there, Charles forwarded animals to Henry in Hoboken, New Jersey, or to clients in Europe. By the late 1870s, Henry was no longer able to stock his Chatham Street store with all their wild cargo. Temporary storage arrangements had been made with the developing Central Park Zoo, but two acres were also acquired in Coney Island.[21] By 1870 the Reiche brothers were worth $100,000 to $150,000; thirteen years later, in 1883, their net worth reached $300,000.[22]

The growth of the Reiches' business can be partially charted by the activity of Adam Forepaugh, a prominent American showman. Forepaugh began in Philadelphia as a horse dealer for streetcar lines. After selling horses to a Philadelphia circus owner in 1864, Forepaugh found himself owner of the show by the next year. Before he embarked on the tenting season of 1867, Forepaugh had bought $35,000 worth of animals from the Reiches. By the 1875 season, Forepaugh's business with the Reiches reached $95,000.[23]

For one animal dealership during a three-year period in the late 1860s, twenty lions, twelve elephants, six giraffes, four Bengal tigers, eight leopards, eight hyenas, twelve ostriches, and two hippopotami generated $112,000 in business, or more than $37,000 per year for large beasts alone. Smaller show

Van Amburgh and Company featured four of the principal attractions in their menagerie, including "The Only Double Horned Rhinoceros in America!" which was imported in 1872. This color poster, probably dating from the late 1870s, advertises an appearance by the menagerie in Gardner, Massachusetts, on June 18. The chromolithographed poster was printed by the Courier Company of Buffalo, New York. (Circus Galleries, John and Mable Ringling Museum of Art, Sarasota, Florida)

beasts as well as monkeys for organ grinders and birds as domestic pets more than doubled the amount.[24]

The prosperity of the Reiche brothers was undoubtedly due in part to the intense competition among American showmen in the number of elephants exhibited. In 1875 P. T. Barnum and his great rival Forepaugh each owned four elephants; by 1878 the Sells brothers' circus boasted seven, the largest number of any show. The next year Forepaugh had twelve elephants, Barnum exhibited eleven, and Cooper and Bailey owned ten (including one that was pregnant and would be the first to give birth in captivity). A total of at least fifty-two circus elephants existed in the United States. In 1880 Forepaugh had eleven pachyderms, but when the large herds of the two Bailey and Barnum shows merged the next year, Forepaugh had to counter with twenty-one of his own elephants in 1882 and twenty-five by 1883.[25]

Competition in numbers of elephants quickly evolved into a contest for the largest elephant when the Barnum show imported Jumbo in 1882. Forepaugh countered with Bolivar (acquired from the Van Amburgh menagerie), and W. W. Cole's circus trumpeted Samson, which it purchased from the Reiche brothers. The elephant wars culminated with a battle to exhibit a white elephant, and Barnum's show secured the first genuine specimen brought to America for its traveling season of 1884. Rivals such as Forepaugh and Cole quickly countered with whitewashed elephants. "[Each declared his] to be the whitest and published convincing medical certificates to prove it," recalled one veteran circus press agent, "but the imposture was so palpable to the public that it was received as a joke. The rival managers were so serious in trying to bolster up their claims,

"THE MOTHER ELEPHANT, 'HEBE,' and her BABY, 'YOUNG AMERICA.'" The first elephant born in captivity, "Young America" weighed 214 pounds and was 35 inches high when delivered in the Philadelphia winter quarters of the Cooper and Bailey circus. *Carte de visite* photograph "from life" by A. W. Rothengatter and Co., Philadelphia, 1880. (Author's collection)

however, as to expend vast sums of money in extra advertising, whose only result was to create distrust. I shall always believe," the agent concluded, "that it required years to overcome this grievous mistake of the circus world."[26]

Besides helping to supply elephants and other animals, the Reiche firm was also involved in operating one of the first aquariums in the United States. P. T. Barnum included an aquarium at his museum in New York by 1860, and soon thereafter he opened an exhibition in Boston which was exclusively an aquarium. Barnum's aquariums were in association with Henry D. Butler, author of *The Family Aquarium*.[27] In 1876 Butler superintended the opening of the New York Aquarium, but this new project was clearly the creation of W. C. Coup and the Reiche firm.

William Cameron Coup was a showman who lured Barnum into starting his famous circus in 1871. The creative Coup withdrew from the circus partnership after five years when he was worth $100,000 or more. Coup then joined with Reiche, and together they invested nearly $60,000 in the New York Aquarium, which opened on October 11, 1876, at Broadway and 35th Street. Six months later another $60,000 was invested in a second aquarium, at Coney Island.[28]

The New York Aquarium for fresh- and saltwater fish was located in a brick building that had a small second floor over the entrance containing a free scientific library and laboratory. The center of the main hall contained a circular tank six feet high and thirty feet in diameter. This tank was intended for a whale that died before it could arrive for the opening, though another whale was soon procured. Nearby was a tank made of Portland cement which contained three seals and a number of turtles. At the far west end was a rocky, stepped grotto for sea lions with a painted backdrop suggesting their natural environment. Along the north and east sides were several large tanks including one sixty-five feet long by six feet high for porpoises and sharks and other large fish. The south wall contained numerous table tanks as well as a fish hatchery.[29]

The aquarium's attractions were an immediate hit. After the first year, a ninety-page guidebook was compiled by the manager, Herman C. Dorner, who had earlier been associated with the Hamburg zoological gardens and its aquarium. In a lengthy article about the study of natural history in America titled "Our Neglected Science," the *New York Times* of June 30, 1871, reported that "no aquarium in the world, except, perhaps, the one at Naples, [could] show so rich a record of tropical species exhibited."[30]

Following a disagreement with Reiche, Coup withdrew from the partnership after one year and returned to the traveling circus business in 1878. His new circus featured a traveling aquarium in charge of "Prof. H. D. Butler," erstwhile superintendent of permanent aquariums for Barnum and the Reiche/Coup partnership.[31] Reiche continued the permanent aquarium, adding various variety acts, operas, and plays (there is clearly something fishy about *Il Trovatore* and *Uncle Tom's Cabin* in this environment) as well as pigeon shows to maintain popular interest. The aquarium closed, however, in 1881, and on April 26 of the same year its tanks, plaster statues, and other remnants were auctioned.[32]

Charles Reiche, a diabetic, died in Europe in 1885; his brother Henry, also diabetic, died in 1887 on a trip to Germany. The firm was succeeded in Germany by the Ruhe family—Paul Ruhe had been one of the Reiches' master hunters. Their "only great competitors," according to former partner Coup, "were the Hagenbecks, of Hamburg. Since the death of the Reiche Brothers, the Hagenbecks . . . almost monopolized the trade, supplying the menageries and zoological gardens of the world."[33]

Unlike the great London animal dealer Jamrach, Carl Hagenbeck was able to penetrate the American market. In 1873 P. T. Barnum visited Vienna to see the International Exposition and traveled to other parts of Europe, including Hamburg. In November he first met the twenty-nine-year-old Carl Hagenbeck and purchased fifteen thousand dollars' worth of elephants, giraffes, and ostriches. By 1877 Hagenbeck and the veteran William Cross of Liverpool were the only two animal dealers designated as foreign agents for P. T. Barnum's circus.[34]

This advertising poster's view of the interior layout of the New York Aquarium, which opened in October 1876 and closed in April 1881, substantially agrees with other published descriptions. The aquarium contained freshwater fish as well as "monster living wonders of the mighty deep from every sea" made possible by "a constant stream of water brought from the Atlantic Ocean & kept in circulation by powerful steam engines at enormous expense." This colorful poster was chromo-lithographed by the Strobridge Lithograph Co., Cincinnati, Ohio, in 1876. (Bella C. Landauer Collection, New-York Historical Society)

Hagenbeck had gained an important foothold in the American market and benefited from the elephant rivalries of the early 1880s. In a 1902 letter to the Ringling brothers, Hagenbeck stated, "In the years from 1875 to 1882 I have sent about 100 Elephants over to the U. States, of which the greater part were sold to the Barnum & the Forepaugh shows."[35]

Hagenbeck was involved in more than animal collecting. His work brought him into contact with cultural groups as exotic to the Europeans as were the animals he imported, and his agents were soon importing Singhalese and other ethnic groups to be presented in Germany and elsewhere. Hagenbeck's huge ethnographic exhibitions—great nineteenth-century folk festivals—were immensely popular in Berlin. Hagenbeck, however, was not the first to collect and exhibit exotic peoples: the prince of Condé kept some reindeer and three Laplanders for four months at his menagerie in France in 1772; in 1822 William Bullock exhibited Laplanders—along with their tools, clothing, and some living reindeer—at his London museum.[36] And the Reiches were also involved in collecting native peoples: though troupes had visited Europe much earlier, they successfully exhibited eleven native American Indians in Germany in 1879.[37]

The Reiches' exhibition, and similar efforts by others, coincided with the rise of anthropology as a discipline in the 1870s and 1880s. In August 1877 the financially ailing Jardin d'Acclimatation in Paris engaged a traveling show of African animals which also included fourteen "Nubians." According to a visitor from the Paris Anthropological Society, which welcomed this first display of humans as enthusiastically as did the mass public, the group belonged "to a foreign merchant whose specialty [was] furnishing interesting specimens to the zoological gardens of Europe." In all likelihood, this presentation was assembled by Carl Hagenbeck, who had just made a large importation of animals and people from the Sudan—his second effort at an integrated animal and human ethnographic exhibit. His first had its debut in September 1874 when several reindeer imported in partnership with a Norwegian were accompanied by some Lapps. Hagenbeck credited the idea for the "first of [his] ethnographic exhibitions" to his friend, animal painter Heinrich Leutemann. Hagenbeck believed that the people in his human displays "behaved just as though they were in their native land" when on exhibit and so fulfilled an educational function.

In Paris the effort at the Jardin in 1877 was so successful that the trend continued for another ten years with gauchos, Fuegians, Galibis, Ceylonese, Kalmouks, American Indians, and Ashanti being paraded alongside the animals regularly on display. The popularity of the Jardin's programs helped influence the organizers of the 1889 Paris Universal Exposition to create avenues of imperial villages in order to show life in all its evolutionary stages. At other world's fairs for the next twenty-five years the importance of people as objects transcended the significance of the manufactured goods on display.[38]

Comparable large-scale efforts in the United States first occurred in 1884. In that year, P. T. Barnum's circus contained a "Museum Department" (generally advertised as the "Ethnological Congress") in a tent separate from and larger than the sideshow tent that contained the usual bearded lady, snake charmer, armless man, and living skeleton. The ethnological congress included a Chinese giant and a Burmese dwarf; other features were a family of American Sioux, five Zulus, eight Sudanese, Asian Indians and Afghans, a troupe of Burmese musicians and priests, and the first white elephant brought to America. The show's extensive menagerie, however, was exhibited in a third and yet larger tent. Ten years later, in 1894, following the fame of similar attractions on the Midway Plaisance of the Chicago World's Fair, the effort was repeated and enlarged to include more than seventy participants. For the season of 1895, a different gathering, nearly as large and equally as diverse, was once again presented by the Barnum & Bailey Circus.[39]

In 1893 Hagenbeck created an animal show for the Chicago World's Columbian Exposition which joined numerous ethnic villages along the Midway Plaisance. Hagenbeck's menagerie display was adjacent to his performing arena seating five thousand spectators; several restaurants and cafés on the two floors and roof catered to visitors. Also part of Hagenbeck's Chicago complex was "an aquarium representing the Indian Ocean with all the wonderful plants, fishes, etc., in the condition they live[d] in." The integration of flora and fauna in the aquarium exhibit presages Hagenbeck's more famous effort that opened at his Stellingen Tierpark in 1907. Much earlier, however, Hagenbeck had experimented with panorama backdrops (as had Reiche and Coup at their New York aquarium in 1876) as part of his efforts toward a bar-less zoological exhibition and received a German patent for the idea in 1896. The 1893 Indian Ocean aquarium and his 1896 panorama can be construed as pioneering efforts to show the relationship of animals and their environment. Although native peoples were not presented by Hagenbeck in Chicago, he did display "a vast number of implements, household goods, arms, etc.," in addition to hunting trophies from the many countries in which he did business.[40]

These various exhibitions were certainly entrepreneurial exploitation that reflected racial and imperialistic attitudes of the nineteenth century, but they also may have indicated a recognition by such animal dealers as Hagenbeck that some ways of life were as endangered as the rare animals being collected. Hagenbeck's and other exhibitors' early educational presentations appear to suggest a similarity to late-twentieth-century concerns about the impact of Westerners on less-developed environments—their peoples, animals, and land—and their gradual disappearance. Although imbued with showmanship, some exhibitors and wild animal dealers sometimes seemed to recognize a relationship between humans and animals, the land, and change.

This understanding is worth investigating, for it could provide us with a useful historical perspective on interpretation and public appeal as well as with new ideas about modern zoos and natural history museums. In *Museums of Influence* (a book that grew out of a report he prepared for the International

Commission of Museums), Kenneth Hudson wonders why that which "would seem to be a natural and inevitable development, the hybridisation of a zoo and a natural history museum, should have been so slow in arriving." He credits Hagenbeck's efforts to display animals in comparatively natural surroundings as the first step in such a direction. Hudson, however, believes that only the Northern Animal Park at Emmen in the Netherlands has made any effort at such integration.[41] Interestingly, like the Hagenbeck zoo, it is largely a private operation.

The story of animal dealers and showmen in the nineteenth century represents a spirit of entrepreneurial adventure. Their story may provide us with fresh ideas of how we might educate the world today about the relationship of humans, their many cultures, the natural world, and the changes we continue to bring to each.

THE ORIGIN
AND DEVELOPMENT
OF AMERICAN ZOOLOGICAL
PARKS TO 1899

*T*he European discovery of the New World by Christopher Columbus in 1492 was a significant event having, among many other effects, a profound influence on the study of natural history and the collection of natural history specimens, particularly animals and plants. This and subsequent explorations coincided with (and further encouraged) the proliferation of Renaissance European menageries and gardens for exotic fauna and flora. The New World was full of animals, plants, and people that were somewhat similar to Old World species but still unknown. Colonial explorers and naturalists made observations of these newly found curiosities and gathered as many specimens as they could for the European collections throughout the sixteenth, seventeenth, and eighteenth centuries.

Beginning in the seventeenth century, many colonists also began to collect natural curiosities. As more Europeans settled in North America, commons, town squares, promenades, and gardens (both public and private) were established throughout the colonies.[1] The first cabinets of curiosities, however, were not established until the eighteenth century by wealthy individuals and by newly formed natural history societies. It was from these early gardens and cabinets that the first American botanical garden and natural history museum emerged.[2]

The collecting and keeping of live animals, as opposed to the collecting and keeping of other naturalia, was a different matter entirely, owing to the complex physical, behavioral, and dietary requirements of the many species and the lack of knowledge about them. Nevertheless, at some point in early colonial times, itinerant animal acts exhibiting individual native species began to appear, and by the 1720s individual exotic species began to be exhibited.[3] The menagerie began to appear after the colonial period, in the 1780s to 1790s, and the traveling menagerie emerged about 1813.[4] These collections remained popular until the Civil War. It was not until the latter part of the nineteenth century that city zoological parks began to appear—about a hundred years after botanical gardens and natural history museums had been established.

In tracing the development of zoological parks in America, this chapter examines the cultural development of the zoological park concept as well as the institutional emergence of the zoological parks, including brief histories of the origins of some of the first zoological parks that were established before 1900.

Colonial America and Its Wild Animal Exhibits

Although American natural history began with the attempt to establish a British colony at Roanoke Island (Virginia) in 1585, it consisted of rudimentary observations and the random gathering of specimens for European collections until well into the eighteenth century. Occasional attempts were made to ship live animal specimens to European menageries, but, for practical reasons, stuffed specimens were usually shipped for exhibit in the more numerous European cabinets.[5] The colonists had little time for wild animal collections. The colonial period was characterized by hard work, frugality, simple pleasures, and the need to establish a new American society in what was an overwhelming and threatening wilderness. The prime ingredients for the development of this kind of cultural institution did not yet exist, namely, the urbanization of a large portion of the population, the reduction of a significant amount of the wilderness, knowledge of and appreciation for nature and animals, the development of a wealthy class, the availability of leisure time among the populace, and an acceptance of the animal exhibit as a suitable form of popular entertainment and education.

Colonial America was primarily rural, with few large towns. By 1790 there were only two towns with 25,000–50,000 residents; most other towns had fewer than 5,000 residents.[6] The upcoming urbanized cultural centers of Philadelphia, Boston, New York, and Charleston were still in their infancy. The wilderness frontier was still at everyone's doorstep, and there was a practical, nationalistic, and religious need to conquer, rather than to conserve, this vast wilderness. Little was known about the wilderness regions or about the animals inhabiting them. The sciences had not developed into the specialized disciplines that we are familiar with today. Knowledge about animals was only one aspect of natural history, which was concerned with observation and basic classification.[7] Ecological, behavioral, and other specialized studies were yet to be considered.

There was, nevertheless, a natural curiosity about the wild beasts of the unexplored, or rarely explored, wilderness regions. At some point, the popular entertainment of the colonists began to include itinerant animal acts. These occurred on occasion when someone would show a bear or some other native species at the local tavern or on the village commons, after which he would pass the hat and move on to the next village. With the presentation of a lion in 1720, these animal exhibits began featuring exotic species brought into the colonies by ship captains returning from foreign ports who wanted to make a little extra money by selling these creatures. Most of these animals were brought into

Boston or New York and were usually exhibited in the New England and New York areas and occasionally in Philadelphia and other areas. The lion was followed by a camel in 1721, a polar bear in 1733, and a leopard along with other animals in 1768.[8] This last group may have been an early menagerie, and there may have been other menageries not known to us at this time.

During this period, the character of the colonies was changing to the extent that, in 1743, Benjamin Franklin felt it was time to establish what was to become the first American scientific society. He made the following announcement, which read in part:

> The first Drudgery of Settling new Colonies, which confines the Attention of People to mere Necessaries, is now pretty well over; and there are many in every Province in Circumstances that set them at Ease, and afford Leisure to cultivate the finer Arts, and improve the common Stock of Knowledge.
>
> That the Subjects of the Correspondence be, All new discovered Plants . . . All new discovered Fossils . . . New Methods of Improving the Breed of useful Animals; Introducing other Sorts from foreign Countries.[9]

It was also during this period, in 1731, that John Bartram's garden in Philadelphia emerged from the many gardens that existed at the time as the first true botanical garden.[10] Natural history museums also began to develop among the many cabinets of natural curiosities of the day; the first was the Charles-Town Museum of the Charles-Town Library Society in Charleston, South Carolina, in 1773, followed by the Peale Museum in Philadelphia in 1784.[11] The menagerie, on the other hand, was still in its infancy, and the idea of a zoological park was still almost a century away.

The Early Republic and Its Menageries

The independence of the American colonies disrupted the cultural and scientific life of the colonies, resulting in the withdrawal of European naturalists and European support. During the early republic period, America focused on nationalistic pursuits. American naturalists filled the void left by the Europeans, developing their own educational programs, research support societies, communication networks, and collections. The effort to survive as a newly independent country, however, was of paramount importance, and the situation necessitated a slow emergence of many cultural activities that did not begin to develop in earnest until the early to mid-1800s. The sciences remained somewhat more international in scope, but they also became more self-reliant during this time.

In the early to mid-1800s, those societies and colleges that were established in the 1700s began to flourish, and new ones became increasingly abundant; expeditions increased and became more scientifically oriented; the number of naturalists increased, as did the quality of their research (helped in part by the increase in the number of jobs that were slowly becoming available in academia, government, and business); societies and publications became more numerous and specialized; and disciplines of natural history developed, with the studies becoming more specialized, precise, and quantifiable. Natural history gave way to the natural sciences: botany, zoology, geology, archaeology, anthropology, and others.[12]

Cultural centers were flourishing, having passed through their precarious periods of establishment and early development, and included Philadelphia, Boston-Salem, New York, Washington-Baltimore-Annapolis, Richmond, and Charleston. These were joined later by Cincinnati, Chicago, St. Louis, and New Orleans, followed by others farther west as the country expanded in that direction.

The attitude of Americans toward their newly settled land, and toward nature in general, developed from the attitudes they brought with them from their European homelands. It was a mixture of fear of the unknown wilderness, a practical need to survive, and a need to cultivate the wilderness "wastelands." It was an attitude of land and natural resource utilization which went unchecked from colonization until the late 1800s.[13]

This attitude also included a continuing natural curiosity about native and exotic animals carried over from colonial times. The early Americans still looked forward to seeing the unusual creatures exhibited by menageries, even though many felt such exhibits were a frivolous waste of time and money. A menagerie consisting of reptiles, birds, and quadrupeds was exhibited in New York in 1781. Another followed in 1789 with a tiger, orangutan, sloth, baboon, buffalo, crocodile, lizards, snakes, and various other creatures. The ostrich was exhibited for the first time in 1794 and the elephant in 1796. Also in 1796, New York saw a menagerie of birds, a seal, and about twenty other animals. A larger menagerie with wolves, monkeys, mongoose, numerous small mammals, and birds was exhibited in New York in the following year.[14] Other collections were exhibited in the Boston and Philadelphia areas, and surely there were many others of which we are unaware (for unless their exhibition was advertised in local newspapers, it is difficult to learn about them).

By 1813 the menagerie was on the road, traveling to many major cities and visiting many of the smaller towns on the way. As time went by, additional exotic species were introduced to the American public: the zebra in 1805, the rhinoceros in 1826, the giraffe in 1837, and the hippopotamus in 1850.[15] The large variety of menageries continued until 1835 when the Zoological Institute absorbed those that were in existence. This control ended in 1837, and from then until the Civil War only two major menageries remained. The circus was coming of age, and many exhibited their own menageries, including ones with trained animals.[16]

The Civil War Period and the Transition to Zoological Parks

The Civil War once again interrupted the cultural life of the newly developing nation, including the newly emerging American sciences. The post-Civil War period was a time of growth and expansion for the country, allowing for the development of many social and economic activities that had begun prior to the war. At the same time, it provided a break with the past, allowing for changes in other social and economic activities. This period was characterized by the shift from rural agriculture to urban industrialization, the loss of wildlife and a concern for conservation, the distancing of increasingly large numbers of the populace from nature, an increased knowledge about nature and animals, and a higher level of education, all of which provided a favorable climate for the development of the zoological park.

The post-Civil War period likewise provided a cultural milieu that favored the transition of the privately owned and operated natural history collection into the professionally administered public institution. During this time, the natural sciences became increasingly specialized and professionalized; popular science increasingly became academic science; open urban spaces became professionally designed parks; gardens increasingly became botanical gardens; cabinets increasingly became natural history museums; and menageries finally became zoological parks.

In particular, the professionalization of the natural sciences and the increasing number of scientific expeditions during the mid- to late 1800s dramatically increased the amount of material being accumulated and the number of species that were becoming known. This encouraged the development of the collections, but at the same time it imposed a financial burden on the individuals who were maintaining the collections. It was also felt that government (at all levels) had no right or authority to be actively involved in the operation of cultural facilities or scientific work. Such things were considered to be private matters, and most collections were therefore maintained by private individuals. Eventually, in the early 1800s, various levels of government began to get involved with the funding of scientific work and collections. Federally sponsored expeditions, the creation of the Smithsonian Institution, state-sponsored geological surveys, and the operation of the first public museum by the city of Charleston are early examples of government involvement.[17]

It gradually became socially and financially acceptable for governments to administer or fund scientific activities and cultural facilities, including zoological parks. With the acceptance of the notions that governments or scientifically oriented civic societies were needed to fund improved menageries, that it was acceptable for governments to administer animal collections for the public good, and that improved menageries could be beneficial cultural institutions, the foundation was set for the establishment of zoological parks in America's communities.

European Zoos and the Launching of Zoos in America

There was no sudden transition to zoological parks. The idea of a zoological park, as distinct from a menagerie, developed gradually starting in Europe, where the menagerie had a long tradition. The Tiergarten Schönbrunn (Vienna) was established in 1752, continuing the royal menagerie established in 1569. The Menagerie du Jardin des Plantes, Muséum National d'Histoire Naturelle (Paris), was established in 1793, continuing the Menagerie Royale de Versailles established in 1665. The Zoological Gardens of the Zoological Society of London were established in 1828, continuing the Tower of London Menagerie established in 1235, itself a continuation of a royal menagerie established about 1100.[18] The Zoological Gardens of London appear to be the first to use the name "Zoological Gardens" and to be established as a grand and extensive collection, as opposed to being just another miscellaneous grouping of caged animals. On paper, from the time of the original idea as conceived by Sir Stamford Raffles (in 1817) until 1827, it was referred to as a zoological collection; in 1827–28 it was referred to as "the gardens" or "the vivarium"; and finally, in 1829, a year after its opening, the name "Zoological Gardens" was used.[19]

It should be noted, however, that the Zoological Gardens of London were also referred to as "the gardens and menagerie" and were still referred to as a menagerie at their centennial.[20] The Paris collection in the Jardin des Plantes is, to this day, known as a menagerie. The use of the term *vivarium* goes back to 1600 and the term *menagerie* to 1712.[21] These were commonly used to designate collections of live animals exhibited according to the then acceptable standards of animal care. It was not until the improved standards of the newer zoological parks contrasted with those of the existing menageries that the latter were considered to be improperly kept collections. The term *zoological garden* evolved with the London Zoo, but without a clear definition of what a "zoological garden" should be and without a clear distinction between it and a "menagerie." In America the word *menagerie* assumed a negative connotation in common usage, implying improperly kept, caged animals, despite its continued use at many well-respected institutions, primarily in Europe.

During the early to mid-1800s, there were many acceptable menageries in Europe, and those in the major European cities were becoming important cultural institutions. They were often visited by Americans who could afford to travel to Europe and who were influential in civic affairs in their American hometowns. This was a time when Americans were still impressed and influenced by European science, technology, and culture. At the same time, they had strong nationalistic feelings and wanted to surpass these European endeavors. They brought back European knowledge and ideas in an effort to improve and Americanize them. One of these cultural ideas was the zoological park.

The first known concern for establishing a zoological park in America was expressed in 1841 by Joel R. Poinsett in his address to the National Institution for the Promotion of Science (although travel accounts of those Americans who visited Europe and the newspaper editorials of the period may reveal earlier concerns). He called for a national institution that would include, among other things, a zoological garden:

> There can be no doubt that a national institution, such as we contemplate, having at its command an observatory, a museum containing collections of all the productions of nature, a botanic and zoological garden, and the necessary apparatus for illustrating every branch of physical science, would attract together men of learning and students from every part of our country, would open new avenues of intelligence throughout the whole of its vast extent, and would contribute largely to disseminate among the people the truths of nature and the light of science.[22]

Interestingly, the effort to develop the Smithsonian Institution, which was under way at the time, did not include a zoological park, even though it was suggested that it should emulate the Jardin des Plantes in Paris, which did have one. The Smithsonian Institution was eventually established in 1846, without any consideration at that time for a zoological park.[23] In 1859 plans were made in Philadelphia to establish one, marking the beginning of the American effort to conceptualize just what a zoological park should be.

The Birth of the American Zoo

Until 1800 Philadelphia was the cultural capital of the early American republic as well as the largest American city. It was the location of the first American scientific society (the American Philosophical Society, 1743), the first botanical garden (the Bartram Botanical Garden, 1731), and the second natural history museum (the Peale Museum, 1784). It was also a city distinctive in its patronage of science and in its civic pride. The antebellum years found Philadelphia with a number of public parks, gardens, museums, circuses, menageries, concerts, theaters, and other cultural entertainments.[24] Thanks to some dedicated naturalists and civic leaders, Philadelphia also found itself with America's first zoological park.

Although menageries existed at the time, some of which were eventually to develop into zoological parks, the Philadelphia Zoological Society was the first to make the effort to establish and open a zoological park. Its large collection of animals housed in several permanent buildings and its management by a professional staff with the support of a community-based zoological society contrasted sharply with the menageries of the day.

A meeting of naturalists and civic-minded citizens was held at the home of William Camac, M.D., leading to the incorporation of the Zoological Society of Philadelphia on March 21, 1859. Dr. Camac requested the meeting, for he had traveled widely in Europe and was well aware of the value of a zoological park to the community. "The object of this corporation shall be the purchase and collection of living wild and other animals, for the purpose of public exhibition at some suitable place in the City of Philadelphia, for the instruction and recreation of the people."[25] In addition, an early report to the society's board of directors stated that it was "the aim of the Managers, not only to afford the public an agreeable resort for rational recreation, but by the extent of their collection, to furnish the greatest facilities for scientific observation."[26]

Although Dr. Camac knew the benefits of a zoo, many others in Philadelphia could not understand the objectives of the society or the benefits to be derived from a collection of animals. This resulted in several failed attempts to raise funds for the new facility. In addition to this, the site assigned to the society in Fairmount Park was in a bad location between two railroads and the Schuylkill River, with no easy access. Because of this, the society may have been reluctant to develop the site for the zoo.[27] Thus, little was accomplished in the first two years, after which the Civil War broke out, diverting everyone's attention to other matters.

In March 1872 a reorganization meeting was held by Dr. Camac, and a new site in Fairmount Park was obtained. A fence was erected around the site and an engineer selected to design the zoo. Funds were raised and work began. Frank J. Thompson, an animal collector, was authorized to obtain animals for the collection and was appointed as the first superintendent of the zoo.[28]

On July 1, 1874, the zoo opened to the public with 212 animals on exhibit, including antelope, lions, zebras, kangaroos, an elephant, a rhinoceros, a tiger, and 50 monkeys. There were also some 674 birds and 8 reptiles.[29] Just two years after its opening, the zoo was joined by the U.S. Centennial Exhibition in Fairmount Park. This helped draw the zoo's largest crowds until recent times, 677,630 visitors (although the centennial exhibit drew 9,910,966 visitors).[30] Thus, while the zoo was popular, it was apparently not the kind of attraction which enjoyed widespread popular appeal. This resulted in several periods of hard times for the zoo. The city of Philadelphia finally recognized its value in 1891 with the first of its financial contributions.[31]

In addition to other accomplishments, the zoo established the first research institution associated with a zoo — the Penrose Research Laboratory in 1901. This facility, along with the New York Zoological Park (established in 1899), would begin a new era in developing the zoological park concept in America.[32]

New York's Central Park Menagerie was one of the menageries in existence at the time the Philadelphia Zoological Garden opened. The menagerie was used as a dumping ground for unwanted pets and unwanted touring carnival animals. They were kept in the basement of the Arsenal Building on Fifth Avenue and in small wire enclosures on the mall near the Casino. By 1865 there were so many animals that they had to be moved to the floor of the Arsenal. The animals were donated as early as 1861 or 1862, but it was not recognized as an official zoo until 1873 when the first formal report of its activities was made to the superintendent of parks. It consisted of a black bear, a pair of Kerry cows, Virginia deer, monkeys, raccoons, foxes, opossums, ducks, swans, pelicans, eagles, and parrots. Even then, however, it remained a menagerie until well past 1900.[33]

In 1868 the Central Park Zoo (the only one established in a city park at the time) sent the Lincoln Park commission in Chicago two pairs of mute swans for display on one of its small ponds, marking the beginning of Chicago's first zoo. Other animal contributions followed, and by July 1873 there were 27 mammals and 48 birds. Thirty-five years later, in 1908, those numbers would increase to 782 birds and 117 species of mammals. Each Chicago park district had its own menagerie, but after these districts consolidated in the 1930s, the collections were discontinued, leaving Chicago with only the Lincoln Park facility.[34]

Victorian gate house with delicate wrought-iron work greets visitors at the entrance to the Philadelphia Zoo (ca. late 1800s). (Zoological Society of Philadelphia)

Panoramic view of the main entrance to the Philadelphia Zoo (ca. early 1900s) shows the Schuylkill River and the city of Philadelphia in the background. (Zoological Society of Philadelphia)

The Zoological Society of Cincinnati was established in 1873. A site was acquired the following year, and the zoo opened to the public on September 18, 1875. The opening collection consisted of an elephant, a tiger, and a hyena but was soon supplemented with many other species. The Cincinnati Zoo had several setbacks and difficulties, and in 1874, before the gates were even open to the public, it was reported that "some errors were to be expected in an enterprise so novel in America as the establishment of a Zoological Garden, but more have been made than can be accounted for from that reason."[35] Nevertheless, by 1891 it was reported that the collection was "the largest and most complete zoological gardens in the country" and was "the only important institution of its kind that [had] neither state nor municipal assistance."[36] Early on, the zoo experimented with bar-less outdoor enclosures, and it is known for having the last passenger pigeon, which died at the zoo in 1914, as well as the last living Carolina parakeet, which died there in 1918.[37]

Bear pits at the Cincinnati Zoological Garden(ca. 1871). The enclosures were modeled after European exhibits. (Cincinnati Zoo archives)

Overview of the Cincinnati Zoological Garden (ca. 1876). The illustration shows the extensive scale of the zoo (sixty-six acres) with a great variety of exhibits and buildings interspersed in a park setting. (Cincinnati Zoo archives; Jon C. Hughes, photographer)

The Buffalo Zoo began in 1875 with a pair of deer that were given to the city by a local businessman who did not know how to take care of them. They grazed in a city park that was a beehive of outdoor activity. When the deer herd increased, it was fenced in, and some cows, sheep, and buffalo were added to the collection. By 1894 the collection was recognized as the Buffalo Zoological Gardens, under the control of the City Parks Commission.[38] In 1895 there were 22 animals in the Buffalo Zoo. That same year Frank J. Thompson, who had been with the Philadelphia Zoological Garden and the Cincinnati Zoo, was hired as the curator of the Buffalo Zoo. New facilities were built, and the collection increased to 138 animals. In 1899 it was reported that what the city needed was "not a menagerie, and a poor one at that, but a Zoological Park—not the penning of a few animals and birds in inadequate fields, houses and cages, but places for them to move about."[39] A steady effort to do this began at the turn of the century.[40]

On April 7, 1876, the Maryland state legislature authorized the Baltimore Park Commission to "form a zoological collection within the limits of Druid Hill Park, by the purchase and collection of living wild, and other animals for the purpose of public exhibition for the instruction and recreation of the people."[41] The collection started with a herd of deer and a flock of sheep. The annual reports of the Park Commission indicate that by the 1890s the Baltimore Zoo contained an extensive collection, including camels, monkeys, cats, an alligator, and a variety of birds.[42]

In September 1882 a site and a herd of deer were donated to the city of Cleveland in order to start a zoo. By 1888 a zoo building and other animal enclosures had been built, and the animal collection consisted of 2 black bears, 2 cougars, a family of raccoons, a pair of foxes, and a colony of prairie dogs. By 1895 the zoological division of the Cleveland Parks Department reported some 321 animals in the collection; however, 200 of these were doves. Among the others were wolves, black bears, elk, deer, a badger, and martens.[43]

The Maturation of the American Zoo

Fifteen years after the Philadelphia Zoological Garden opened to the public, the National Zoological Park was established in 1889. During this period the country had, for the most part, recovered from the Civil War and was in a new era of municipal improvement and national pride. The first national park was established at Yellowstone in 1872, and the conservation of our natural environment was becoming a national issue. The major cities were becoming more urbanized, and the public park was becoming increasingly popular for social activities.

The zoological park concept fit into this changing cultural setting very well and was about to leave its embryonic stage and mature into a cultural institution that civic leaders and citizens could point to with pride. Although the twelve zoos in existence at this time were, for the most part, still menageries, people were beginning to recognize their recreational and educational value, and the zoological park was becoming a symbol of America's greatness. The National Zoological Park was the first of the zoos to exemplify these values.[44]

During this period, some of those who looked forward to the coming of the zoological park felt they should have open, naturalistic habitat enclosures, natural diets, and scientific management. This was the intention of several zoos, but because of financial and other constraints it was not possible to meet these goals. Most of the established zoos were criticized for being glorified menageries rather than zoological parks.[45] This was rectified with the creation of the country's National Zoo, which was to set a new direction in the development of the zoological parks.

Joel R. Poinsett first called for a national zoological garden in 1841. P. T. Barnum wanted to turn his museum into a national institution and was in favor of a national zoological park, first in New York (1865) and later in Washington (1888).[46] On June 21, 1870, the Washington Zoological Society was established; however, nothing apparently came of this society or its efforts to establish a zoo.[47] Many considered it a disgrace that our capital did not have a zoological collection, but many others felt it was not the federal government's responsibility to finance such an institution. Nor was a collection of living animals considered an appropriate function of the Smithsonian Institution, which was established in 1846 with a museum of natural history but not a zoo. A museum was more in keeping with the needs of the naturalists, the public, and Congress at the time.

In 1887 the Smithsonian's Museum of Natural History established the Department of Living Animals "in order to afford to the taxidermists an opportunity of observing the habits and positions of the various species, with a view to using the knowledge thus acquired in the mounting of skins for the exhibition series of mammals."[48] William T. Hornaday, the chief taxidermist, was appointed as the curator in 1888. The living collection became a popular attraction and resulted in the donation of many additional animals. By March 1888 the collection had nearly doubled in size and required a full-time keeper to take care of the animals. Although the collection was considered to be a menagerie, its popularity reinforced the contention that Washington needed a zoological park. This point was pushed by Hornaday and others.

On March 2, 1889, Congress passed a bill authorizing the establishment of a national zoological park "for the advancement of science and the instruction and recreation of the people."[49] As described by Smithsonian secretary S. P. Langley, the new zoo would serve as a "refuge for the vanishing races of the continent"; it would retain much of its natural beauty while, at the same time, providing spacious outdoor exhibits. Only 25 percent of the new zoo's 166 acres would be allowed public access—most of the space would be reserved for the animals.[50] Unfortunately, not all of these lofty ideals were achieved; nevertheless, they did serve as the defining concept of what a zoological park should be at the end of the nineteenth century.

Having been overlooked for the position of director of the new zoo, Hornaday resigned in 1890 and went into private business until 1896, when he became the director of the New York Zoological Park and made it a rival with "a reputation that the National Zoo long coveted."[51] The National Zoo opened to the public on April 30, 1891.[52]

The year 1889 also saw the establishment of a zoo in Atlanta. On March 28, 1889, G. V. Gress, a prominent businessman, purchased the animals of a bankrupt circus and immediately announced his intention to donate the collection to the city for the establishment of a zoo in L. P. Grant Park. The city accepted the collection, which consisted of two lions, two cougars, two wild cats, monkeys, a black bear, a raccoon, a hyena, a jaguar, a Mexican hog, a Bactrian camel, a dromedary camel, an elk, and a gazelle. A menagerie building was constructed which served its purpose for seventy years.[53]

The San Francisco Zoo had its beginnings in 1889 as well, when a grizzly bear was exhibited in Golden Gate Park. In 1928 the zoo moved to its present

A Washington, D.C., public school group looks at parrots in front of the National Zoo's former lion house in 1899. The building was torn down in the early 1970s and has been replaced by a large, open-air, naturalistic facility. (Frances Johnston)

Schoolchildren in 1899 inspect the first bison at the National Zoo. Today, nearly one hundred years later, school groups continue to flock to the zoo. (Frances Johnston)

site. The collection consisted of two zebras, a Cape buffalo, a Barbary sheep, five rhesus macaques, two spider monkeys, and three elephants.[54]

The Denver Zoo had a modest beginning as well when a black bear cub was given to the mayor in 1896. He directed that the bear be taken to the City Park and tethered to the haystack there. This fellow was soon joined by other native species such as eagles, wolves, and deer, which became the nucleus of what was to become the city's zoo.[55]

In 1896 the city of New York established and managed an aquarium in what had been originally built as a fort in 1807. In 1902 the aquarium was turned over to the New York Zoological Society. The collection consisted of an extensive selection of native freshwater and marine fishes, as well as amphibians and aquatic reptiles. Aquatic mammals such as harbor seals, West Indian seals, and manatees were also exhibited. In 1897 the aquarium also exhibited white whales. Although the building served many functions, it is not known why it was decided to convert it to an aquarium; perhaps it was because it was near the ocean and in a public amusement area. There were also other aquariums in the city. Whatever the reasons, the New York Zoological Society now had an aquarium to complement its newly developed zoological park—a unique feature that few zoological parks were able to duplicate.[56]

The Pittsburgh Zoo opened to the public on June 14, 1898, as the Highland Park Zoo. The effort to start the zoo actually began in 1895, and when the zoo opened, it already had a building to hold the elephants, lions, tigers, antelope, zebras, and cougars that had been bought and donated to the city.[57]

In 1899 the Toledo Zoo acquired animals for its opening the next year. The collection consisted of a woodchuck, a wolf, a pair of foxes, raccoons, rabbits, alligators, a badger, an opossum, and an assortment of birds.[58]

The Realization of the Modern American Zoo

In 1899 ten years had passed since the National Zoological Park started moving the zoological park concept in its new direction and established the American zoological park as a mature institution. The National Zoological Park was established for scientific purposes as well as for education and recreation; however, it was not until the New York Zoological Park began its work that these purposes, as well as wildlife conservation purposes, were actively pursued.

As early as 1884, a report on public parks stated that "no park system [could] be regarded as complete without suitable tracts for botanical and zoological gardens" and that "a park system that failed to include a zoological garden would be wanting in one of the most essential requisites."[59] One might wonder, however, why these calls to establish such a facility in one of the country's major cities had not occurred well before this time. Perhaps the answer was that the city was considered a center of crass commercialism and lacked cultural interest in such matters.[60] Whatever the reason, the city's civic leaders, concerned citizens, and those sportsmen who were concerned with the preservation of the country's wildlife came together and established a zoo with priorities that were, for the first time, considered an imperative function of a zoo—research and conservation. After some years in private business and thoroughly disappointed with zoos, based on his experience with the National Zoo, William T. Hornaday was convinced to return to zoo administration and subsequently became the first director of the New York Zoological Park. He was finally able to do what he could not do in Washington, and in the process he set the pace for all other zoos.

The New York Zoological Society was incorporated on April 26, 1895, "to establish and maintain in [New York] a zoological garden for the purpose of encouraging and advancing the study of zoology, original researches in the same and kindred subjects, and of furnishing instruction and recreation to the people."[61] The society's first annual report stated that its objectives were "the establishment of a free zoological park containing collections of North American and exotic animals, for the benefit and enjoyment of the general public, the zoologist, the sportsman and every lover of nature"; "the systematic encouragement of interest in animal life, or zoology, amongst all classes of the people, and the promotion of zoological science in general"; and "cooperation with other organizations in the preservation of the native animals of North America, and encouragement of the growing sentiment against their wanton destruction."[62] The New York Zoological Park was opened to the public on November 8, 1899, with a collection of 843 specimens representing 157 species.

With the establishment of the New York Zoological Park in 1899 and the Penrose Research Laboratory (at the Philadelphia Zoological Garden) in 1901, the full range of objectives of a modern zoological park was achieved: recreation, education, scientific research, and conservation. The New York Zoological Park and the developing collections at Philadelphia, Cincinnati, and Washington were shaping the idea of the modern zoo.

By 1900 some 32 zoological parks had been established in the United States. After the turn of the century, numerous other zoos were established. Although many of these, as well as those already in existence, were not much better than the earlier menageries, they now had modern models to emulate. This effort was given a boost in the 1930s with the help of the Works Progress Administration (WPA), which assisted in the needed renovation of a number of zoos. Unfortunately, zoos had to endure setbacks in the 1940s as a result of World War II, but they bounced back with modern buildings in the 1950s. Beginning in the 1960s, zoos developed serious conservation goals, established zoo libraries, constructed state-of-the-art research and veterinary facilities, and created a larger variety of professional and administrative positions. They began to build truly naturalistic, ecologically oriented exhibits in the 1970s and 1980s (as was originally anticipated in the 1800s).[63]

The concept of the zoological park continues to evolve as it merges with the botanical garden, the natural history museum, and the public park. The establishment of the Arizona–Sonora Desert Museum in 1952 marks the beginning of

Baird Court, Bronx Zoo, June 1910. The Primate House is at the left, the Lion House is at the right, and the sea lion pool is in the foreground. (New York Zoological Society photo archives)

Primate House interior, Bronx Zoo, February 1906. (New York Zoological Society photo archives)

The old Bird House, Bronx Zoo, July 1907. (New York Zoological Society photo archives)

The Birth of the American Zoo	Philadelphia Zoological Garden, 1859/1874
	Central Park Zoo, 1861/1873
	Lincoln Park Zoological Gardens, 1868
	Roger Williams Park Zoo, 1872
	National Aquarium (Washington, D.C.), 1873
	Cincinnati Zoo, 1873
	Buffalo Zoological Gardens, 1875
	Ross Park Zoo, 1875
	Baltimore Zoo, 1876
	Cleveland Metroparks Zoological Park, 1882
	Metro Washington Park Zoo, 1887
	Dallas Zoo, 1888
The Maturation of the American Zoo	National Zoological Park, 1889
	Atlanta Zoological Park, 1889
	San Francisco Zoological Gardens, 1889
	St. Louis Zoological Park, 1890
	Dickerson Park Zoo, 1890
	John Ball Zoological Garden, 1891
	Miller Park Zoo, 1891
	Milwaukee County Zoological Gardens, 1892
	New Bedford Zoo/Buttonwood Park Zoo, 1892/1894
	Prospect Park Zoo, 1893
	St. Augustine Alligator Farm, 1893
	Seneca Park Zoo, 1894
	Denver Zoological Gardens, 1896
	New York Aquarium, 1896
	Como Zoo, 1897
	Pittsburgh Zoo, 1898
	Henry Doorly Zoo, 1898
	Alameda Park Zoo, 1898
	Toledo Zoological Gardens, 1899
The Realization of the Modern American Zoo	New York Zoological Park, 1899

this new direction toward an ecological zoo-garden-museum. There is also a merging of *ex situ* (captive) conservation programs and *in situ* (field) conservation programs. The changing of the New York Zoological Park to the Wildlife Conservation Park and the New York Aquarium to the Aquarium for Wildlife Conservation in 1993 marks the formal recognition of this emphasis. Perhaps the 1990s will go even further and bring to a close the era of the zoological park, mere menageries themselves when compared with the upcoming biopark that integrates animals, plants, environmental aspects, and human relationships into a holistic presentation.[64]

THE NATIONAL ZOOLOGICAL PARK

"City of Refuge" or Zoo?

*I*n 1889 the National Zoological Park was established in Washington "for the advancement of science and the instruction and recreation of the people."[1] Prior to this founding, American conceptions of wild animals were derived from three different types of experience: the hunt, the animal show, and previously established zoological gardens.

European settlement on the continent had involved the displacement, exploitation, and extermination of native peoples and indigenous animals. Some American animals were merely in the way; they were considered nuisances or dangers to be eliminated. Others were economic boons—sources of food or valuable fur. In either case, the meaning of the wild animal was the same: the animal was the object of the hunt. It was the animal dead that was desired. Whatever pleasure was derived came from economic gain, personal security, or the thrill of the chase. The wild animal was hardly valued in and of itself.

Yet there were other kinds of animals—strange, exotic beasts that inhabited other continents. From the early eighteenth century, Americans had seen some of these wondrous creatures in traveling menageries that toured cities and towns. By the mid-nineteenth century, the great circuses and animal shows were gathering elephants, lions, tigers, and giraffes in large numbers to entrance, surprise, frighten, and titillate paying customers. Beyond their strangeness and beauty, the attraction of the wild animals in these shows was twofold: either the animals were made to perform in ways that defied their wildness, or they were oddities of nature, analogous to the freaks with whom they often appeared.[2] The animals were sources of curiosity and pleasure. Some, such as the elephant Jumbo, were well-known individuals. Yet the other side of fascination was contempt for the menagerie animals, perhaps the necessary tribute to a morality that scorned such amusement.

Well-traveled or more cosmopolitan Americans experienced wild animals through a distinctive medium, the zoological garden. The display of wild animals in a public garden designed for recreation and enlightenment came with the removal of the deposed Louis XVI's collection to the Jardin des Plantes in Paris during the French Revolution and with the opening of the London Zoo in 1828. By 1870 Americans could see the great German zoological gardens of Berlin, Frankfurt, Hanover, and Cologne constructed on a monumental scale. In such cities as New York, Cincinnati, and Philadelphia, Americans were attempting to re-create these European institutions within the limits of their means.[3] Such zoos must have seemed like the parks in which they were set, delightful additions to urban life and sources of healthful recreation to the population. The animals themselves were almost secondary, elements of the landscape that lent interest and charm and gave an excuse for decorative garden architecture.

The Central Park Zoo in New York, founded in 1861, consisted of caged specimens of exotica gathered in a small section of the great park along Fifth Avenue. While it may have been seen as a refined version of an animal show, its placement and display suggested the Jardin des Plantes in Paris. The Cincinnati Zoo, organized a decade later, was different. Resembling the recently developed zoo in Hamburg, Germany, it sheltered its animals in decorative buildings that were spaced generously along softly rolling grounds. Well-dressed adults could stroll on sunlit days along its walkways into its animal buildings or its festive restaurant.[4]

Philadelphia undertook a similar enterprise; however, here the model was the London Zoo with its scientific society and educational goals. The buildings in Philadelphia were designed in a romantic mode, communicating a sense of mystery and even foreboding. The animal kingdom took on an awesome aspect enhanced by impressive theatrical buildings within a wooded landscape. It was still a place for adult recreation, but one of melodrama rather than of comedy.[5]

Beginning in 1888, the Smithsonian Institution in Washington, D.C., attempted to establish an alternative way in which the public might experience wild animals. Established by the unanticipated bequest of an English scientist to lead to "the increase and diffusion of knowledge," the Smithsonian was, for several decades, largely a research institution. Its second secretary, Spencer F. Baird, shifted emphasis from research to the development of the National Museum and to the gathering of a collection of natural history and ethnography artifacts.[6] Under Baird's assistant secretary, George Brown Goode, a prominent ichthyologist, the National Museum grew from two hundred thousand specimens to more than three million.[7] Although the size of the holdings is notable, it is the understanding behind the acquisition which is significant. According to Goode, the museum had a dual purpose: it contributed to the culture of the public "through the display of attractive exhibition series, well planned, complete, and thoroughly labeled," and its study series served as the material base for original investigation. As an adjunct to science, the museum both stimulated research and was a depository of record in which would be placed specimens "upon which critical studies have been made in the past" and those that were "landmarks for past stages in the history of man and nature."[8] As a naturalist as well as museum administrator, Goode saw his specimens as important to science for both pure research and potential economic value.[9]

In 1882 Goode hired William Temple Hornaday as chief taxidermist for the growing exhibition of mammals. Hornaday was a young man who distinguished himself on hunting expeditions in the Caribbean, South America, and Asia and who was currently achieving recognition for his naturalistic treatment of specimens and his effort to place them in habitat settings. Once in Washington, Hornaday turned his attention to large, North American mammals. The absence of adequate specimens on exhibit provided him with the opportunity to travel to the Rocky Mountain West in search of bison. What he saw changed the direction of his career and had a lasting impact on institutions concerned with the keeping and preservation of wild animals.[10]

The millions of bison that once roamed the continent were gone, systematically butchered. Hornaday's expedition, financed by the Smithsonian, required hard months of hunting to net twenty-five of the last few hundred bison remaining on earth. The finest specimens were mounted in a splendid habitat grouping for the National Museum. Upon his return, Hornaday, the hunter, naturalist, and social critic, wrote his tribute, "The Extermination of the American Bison," published in 1889. The buffalo, once the most numerous quadruped in the world, was about to become extinct, lost to science and to humanity through wasteful human slaughter. Hornaday asked, "With such a lesson before our eyes . . . who will dare to say that there will be elk, moose, caribou, mountain sheep, mountain goat, or black tailed deer left alive in the United States in a wild state fifty years from this date, or even twenty-five?"[11]

George Brown Goode, Hornaday's superior, obviously approved of his taxidermist's work. Hornaday's expeditions to the West had also been sanctioned by Baird, toward whom Hornaday felt great admiration and gratitude.[12] With Baird's death in November 1887, Samuel Pierpont Langley, an astronomer, became secretary of the Smithsonian. Under Langley's supervision, Hornaday carried out his plan for a department of living animals under the National Museum. No longer would the animals that had served as taxidermists' models be killed or shipped away to zoos in Philadelphia or New York. They would now be maintained on the Smithsonian grounds and dedicated to the purposes of science and preservation. The Smithsonian became committed to extending its animal collections from the dead to the living.

Hornaday's cause became Langley's. The Smithsonian began its campaign for an area in Washington to be set aside for "a home and a city of refuge for the vanishing races of the continent," a national zoological park.[13] Hornaday was commissioned to survey the land in the Rock Creek area for a possible site. He also began to solicit gifts or loans of bison from private collectors.

It took two sessions of Congress, 1888 and 1889, to authorize a commission to establish a zoological park along Rock Creek. Strenuous debate ensued in the following two years about the method and amount of funding. The arguments

These bison, as well as other
North American animals,
were on view to the public
near the Smithsonian Castle
in 1888. This collection
was moved in 1891 to Rock
Creek valley, where these animals became the first residents of the National Zoo.
(Smithsonian Archives)

William Temple Hornaday,
chief taxidermist at the
Smithsonian's U.S. National
Museum, leads an endan-
gered bison calf on the Mall,
near the Smithsonian Castle,
in 1886. Hornaday is considered the guiding spirit
behind the creation of the
National Zoo. (Smithsonian
Archives)

Smithsonian Institution officials discuss plans for the
National Zoological Park in
1890, one year after its establishment by Congress. Pictured among the men are the
zoo's acting superintendent,
William Hornaday (*third
from right*), and the landscape architect Frederick
Law Olmsted (*in the light
suit*), famous for his designs
of the U.S. Capitol grounds
and New York City's Central
Park. (National Zoological
Park and Smithsonian
Archives)

were stable, representing clear alternative understandings about the meaning of wild animals and their value in captivity. In Congress, certain senators and representatives became spokesmen for the Smithsonian Institution. Under its aegis was to be established a national scientific institution in which threatened North American species were to be sheltered in the hope that they would breed. In light of the federal government's involvement, it was therefore fitting that it be the only source of funds: it should allocate moneys for the acquisition of a collection including its appropriate housing and maintenance. There was no discernable difference between the collection of living animals in a national zoological park and the stuffed dead ones in the National Museum. The goal was the same, the advancement of science; the authority was the same, the Smithsonian; thus the means of support should be the same, the federal government.[14] Although the national zoo would have a special relationship with Yellowstone National Park for a source of animals, it was important that the zoo be located in Washington to be available to scientists for research.[15] (The Smithsonian's further hope that the zoo might serve "as a constant object lesson under the eyes of the legislature,"[16] a living lobby near the Capitol to assert the value of wildlife and the need for its preservation, was discreetly never stated on the floor.) Such were the arguments that won the Senate in the years of debate, 1888–91.

Although there was strong opposition in the House, a majority did not oppose a zoo but saw it as a local pleasure ground instead of a scientific facility and therefore not a proper way to carry out James Smithson's bequest. One congressman declared, "The Government is being compelled by Congress to provide a 'zoo' for the city of Washington. . . . Every other city has to provide its own 'zoo.'"[17] When the Smithsonian's goals of preservation and science were recalled, they were labeled ideals, and it was argued that those who would actually use the park would be the residents of Washington, who would visit it for the purpose of enjoyment.[18] Under the Organic Act of 1878, District of Columbia public works were to be voted on by Congress and to be supported equally through federal funds and District taxes.

These were the arguments that determined congressional action. The Senate maintained the Smithsonian position stoutly at the time of the founding of the park and during the initial debates over funding. But the House refused to compromise its stand, and, to allow any zoo at all, the Senate had to accept the House's terms. By 1891 it was clear that any opposition was futile and that the form of the zoo was set. It was equally clear that the House would be most parsimonious with public money.

Until then, Langley seems to have believed that the connection with the District of Columbia was merely a temporary arrangement. His plans for the park reflected these hopes. The National Zoological Park was to be unlike any zoological garden existing in America. In contrast with Cincinnati Zoo's 36 acres and Philadelphia Zoo's 40 acres, the National Zoo would hold more than 166 acres "in the picturesque valley of Rock Creek. . . . Here not only the wild goat, the mountain sheep and their congeners would find the rocky cliffs which are their natural home, but the beavers brooks in which to build their dams; the buffalo places of seclusion in which to breed and replenish their dying race; [and] aquatic birds and beasts their natural home."[19] Only a small section of 37 to 40 acres was to be open to the public; the rest was to be a preserve area where bison, elk, and other North American mammals would live and breed relatively free from the presence of humans.[20] Within this reserved area would be space for scientific research, for the animals' medical care, and for the necessary administration.[21]

Secretary Langley hired Frederick Law Olmsted to help him preserve and enhance the natural features of the park and to help him disguise the touch where the human hand was necessary to protect animals or people or to provide facilities for them. Langley instructed that the accidental quality of the park be maintained. Natural ravines were to be respected, there were to be no straight lines, and the fencing was to be disguised.[22]

In terms of the American context, Langley's plan for the park in its reservation of areas for breeding and in its naturalistic treatment is an innovation; however, when one looks across the Atlantic at private estates, there are clear precedents. Langley differed from his European contemporaries in combining the game preserve with the zoological garden. He understood that from certain points the public would be able to view the herds of American mammals. But, in addition, the National Zoological Park would have a section where a collection of native and foreign animals would be on display in the more traditional fashion.[23] Since it was generally not expected that the zoo would have significant funds for purchase, the plan was to use the products of successful breeding as currency: young bison or young elk might be exchanged for kangaroos or ostriches.[24] This dual nature of the National Zoological Park was an extension into the area of living animals of Goode's understanding of a museum: through its study series and facilities it was to serve science; through its exhibition series it was to offer culture to the public.

The principle governing the design of the animal houses followed the nineteenth-century notion of association. Langley argued that the buildings should suggest their animal inhabitants' "habitat"; however, he did not mean that the structures should resemble the desert plains or the Rocky Mountains. Rather, he meant that they should assume the shape of the human dwellings in the area. The buffalo house thus took the form of a log cabin. Langley was following the European tradition of zoo architecture that placed elephants in Indian temples, an effort at scene painting.[25]

That Langley's initial park design was an extension of the National Museum of the Smithsonian became clear in 1891 when the Senate failed to convince the House to change the form of appropriation for the zoo. This time those for whom animal collections conjured up circuses and menageries had their say. The appropriation regarded as absolutely necessary for park development and animal maintenance was halved, and it was expressly forbidden that any money be spent for the acquisition of animals.[26] Reality struck Langley, and on March 4, 1891, he drafted a memorandum to the Smithsonian regents which would critically revise his plan for the National Zoological Park. The measure just passed had "the practical result of substituting for the scientific and national park (with subordinate features for recreation) . . . a local pleasure ground and menagerie." There could be no animal preserves. Of the one-half to two-thirds of the park hitherto reserved, all, except that essential to animal protection, had to be thrown open to the public. Money would have to be diverted from the building of permanent structures to cut roads, footpaths, and entrances into the park. The carnivora house would have to be sized down and no houses built as planned for birds, monkeys, or reptiles. The force of employees and watchmen was to be reduced. Langley argued that whatever money could be saved should be spent for the acquisition of "interesting animals" before the congressional prohibition on purchase went into effect. Even if the growth of the collection presented difficulties in housing and feeding, additions were vital, according to Langley: "Gifts can hardly be expected to come in to any extent until the collection is so impressive as to excite the enthusiasm of people who have such things to give." The very limited appropriation of fifty thousand dollars was to be spent "with the feeling that public approval of and interest in the work [had to] be engaged as the most efficient agent in securing an increase in the annual appropriation."[27]

The Smithsonian regents accepted Langley's memorandum, and it thus became "a general rule of action."[28] Gone were the elements of the National Zoological Park that made it different from a traditional zoological garden. Not only was the public to have access to the entire park, but it should now be able to view the usual set of zoo animals. Downplaying its former concern with North American mammals, the National Zoological Park attempted to get those wild animals that defined a zoo in the public mind: lions, tigers, zebras, swans, polar bears. Its finest early animal was the Indian elephant Dunk, given with a companion, Golddust, by James E. Cooper, proprietor of the Adam Forepaugh shows.[29] During the winter season of 1893–94, the park accepted on loan seventy-three animals from the Forepaugh shows. This generated great popular interest, bringing to the park shortly after their arrival almost thirty thousand visitors on a single Sunday.[30] What the zoo wanted was "interesting animals" that would be pleasing to the public.[31] The study series was forced to give way to the exhibition series.

Whether Hornaday would have forwarded or resisted these changes cannot be known, for by 1891 he was in Buffalo, New York, engaged in real estate ventures. He had felt forced to resign in May 1890, when Langley made it clear that the zoo was to be Langley's park, not Hornaday's.[32] Hornaday felt he could not remain as a "Superintendent of buildings and labor, or a head keeper," reporting to an assistant secretary and powerless to shape development of the zoo, which was to him "as clear as the noonday."[33] The substantive issues between these two powerful men may never be known. Undoubtedly, there was the difficult adjustment for Hornaday, passionately devoted as he was to Baird, when his project was appropriated by the formal, stern Langley, determined to gain authority over his subordinates.[34] Hornaday, too, may have been disheartened by the actions of Congress, for he fully understood the implications of the District tax from the beginning and lobbied hard against it.[35] Finally, whatever his intentions in entering business in New York State, Hornaday was aware that a "big scheme" for a zoological park was afoot in New York City.[36]

Langley acted quickly to fill Hornaday's position. He appointed as acting manager of the National Zoological Park Frank Baker, a specialist in comparative anatomy at the National Museum and professor of anatomy at Georgetown University. Langley expected that Baker would serve only a short time and that he would be more responsible for supervising the construction of buildings than for the actual care of animals.[37] Langley severely restricted Baker's ability to act. Baker could make no move beyond day-to-day management without written authorization, and he had only a very limited ability even to correspond in his own name.[38] Langley's domination of his acting manager was unremitting.[39] Unlike Hornaday, however, Baker bore up, learned to satisfy Langley's demands, and little by little became an efficient bureaucrat, asserting authority over his own staff.[40]

Baker went on to become the zoo's superintendent, a title that he received on November 23, 1893, and was to hold until 1916.[41] Increasingly, he worked at the park rather than at the Smithsonian building on the Mall. He visited zoos in Cincinnati and Philadelphia, was surprised by their success, and learned from their directors.[42] Neither he nor Langley had known anything about the keeping of wild animals in captivity.[43] They were rank amateurs who had to inquire about the kind of enclosures and appropriate diets for specific animals, as well as the effects of climate on them. In 1891 they hired William H. Blackburne as head keeper. He brought them his experience with animals gained in twelve years with Barnum and Bailey. Baker learned quickly, and as he did, he came to have a sense of the National Zoological Park quite different from that of Langley. Baker wanted a place that would please the public immediately. He was concerned with public safety and convenience rather than with preserving naturalistic detail. Even before Langley's memorandum, Baker had wanted the zoo to get as many animals as possible—including the foreign ones the public expected. He feared that when the National Museum's Department of Living Animals was moved to the park, it would appear so small as to "make the whole scheme seem abortive and ridiculous."[44] Thus the changes in design that Langley outlined in May 1891, out of a sense of crisis, must have seemed to Baker sweetly reasonable.

Increasingly, as the National Zoological Park became Baker's, it lost its distinctive qualities to become a modest zoological garden. By 1895, 520 specimens lived on the developed northeasterly segment of approximately forty acres: there was a heated animal house; there were paddocks and barns for bison and elk; there were enclosures and thatched-roof shelters for deer and llama; and there were bear yards, an elephant barn, a prairie dog yard, a waterfowl pond, and a beaver valley. In contrast with the half dozen or so substantial buildings evenly spaced throughout Philadelphia's or Cincinnati's parks, Washington's zoo had only one solid, permanent structure.[45]

The National Zoological Park did have special potential, however. Because it was free, people enjoyed it during all phases of the business cycle. The support from federal and local funds meant the zoo would never face the hard times, even

William H. Blackburne, the National Zoo's first head keeper, gives a young camel a drink in 1893. From the day he arrived at the zoo on January 29, 1891, until his retirement fifty-two years later, Blackburne never took a vacation or a day of sick leave. (Smithsonian Archives)

Easter at the National Zoo (ca. 1910). Crowds of visitors still flock to the zoo on this spring holiday. The zoo's former lion house is in the background at the top of Lion-Tiger Hill. (Smithsonian Archives)

bankruptcy, of its better-equipped counterparts. And, while only occasionally has Congress been generous, the National Zoological Park has been able to improve and build steadily; thus it has been able to incorporate changing concepts rather than remaining essentially fixed by an early plan.

Despite his 1891 memorandum, Langley remained, as long as he lived, a force opposing the National Zoological Park's becoming a traditional zoological garden. He continued to fight for naturalistic detail, for emphasis on North American mammals, and for increasing the size of the animals' paddocks.[46] In his annual reports as secretary of the Smithsonian, he attempted to disguise the extent to which the National Zoological Park had come to resemble other zoos. He instructed Baker to "give prominence to the native races, keeping the others quite subordinate,"[47] to strike out the name "tropical house" whenever it appeared (despite the fact that that was what it was), and to reject pictures showing animals visibly constrained for those of elk and bison in which the fencing of the enclosures did not show.[48]

In 1895 Langley still recalled the hopes of 1889, hopes kept alive by the magnificent setting along Rock Creek.[49] Alongside elements of a traditional zoological garden, what he had originally proposed for the National Zoological Park was an alternative to the ways the American public experienced wild animals. The animal was to be rescued from the hunter and given a protected area where it might replenish its stock. Here, the animal's primary value was not to be exhibited either as entertainment in an animal show or as part of a decorative setting. Its purpose was to have been merely to exist and breed—and only incidentally to be seen. The reason the National Zoo was to be located in Washington clarified its secondary purposes: the collection of animals would be available to scientists, and the animals could serve as a standing lobby for game protection. In this early effort at preservation of wildlife, the animal was important for its place in the natural order and for its value as an economic resource. Neither required emphasis on the display of the animal to the public.

Langley's plan for the zoo to be a refuge and breeding facility was not to be realized—at least not until 1974 when the zoo's 3,100-acre Conservation and Research Center in Front Royal, Virginia, was obtained. Saddled with a special relationship to the District and its taxpayers, the National Zoological Park assumed the form of a traditional zoological garden. Limited by the meager support of a reluctant Congress, it was only a minor affair, well below the scale of other zoological gardens in the United States. Yet, ironically, it may have been the very stinginess of Congress that salvaged something of Langley's original plan. Though there were not the large herds of bison and elk breeding upon it, the land itself was left relatively unchanged, its wildness and beauty intact.

EPILOGUE

*T*ransition, a word so descriptive of our present society, is the
appropriate word for the end of the nineteenth century and the
beginning of the twentieth. The world was changing, trying to
cope with new philosophies, artistic and musical styles, political systems, industries, inventions, technologies, transportation means, and communications
methods. Telephones, automobiles, phonographs, electric lights, airplanes, ma-
chine guns, tanks, communism, nationalism, aggressive capitalism, and oil, steel,
and railroad tycoons—all competed for attention on the world stage. New per-
spectives and new ideas swirled throughout this milieu and began to shape the
new world of the twentieth century.

The international zoo community was also caught up in the push and pull of
this transition. At the turn of the century, the modus operandi of established
zoological parks was being challenged by new ideas on the purpose and presentation of animals.

This book has described the emergence of public zoos in the nineteenth
century from their origins in the menageries of the royal and wealthy of previous
centuries. By the late 1800s, the public zoological park or garden was a flourishing concept and a worldwide reality. No country of any status could be without
one. Citizens in major cities coveted the thought of having the best museums, art
galleries, opera house, and zoological park of any city on the globe. As with the
other cultural institutions, the premier zoo symbolized the premier nation or
empire. The world's best zoo by definition had to exhibit the largest number of
exotic species (especially the rarest or most recently discovered). Zoo animal
collections at the time, it seems, were viewed as examples of "mother nature's"
works of art. Obtaining a newly discovered okapi was equivalent to the national
art gallery's acquiring an original da Vinci. Collecting exotic animals became
analogous to collecting rare or heretofore inaccessible art.

The concept of zoos as art galleries spilled over into the twentieth century.
Even animal buildings and exhibit designs became works of art—architectural
gems, as attested by some of the photos in this volume. Exhibit and building
designs favored the architecture of the foreign lands from which the animals

came: thatched huts, Indian and Egyptian temples, pagodas, rustic lodges, and so on. Animal collecting expeditions to Africa, Asia, and South America occurred regularly and were well financed (wealthy industrialists were particularly enthusiastic backers). However, on the fringe of this national self-glorification and competitiveness that influenced zoo operations, a modern, twentieth-century vision was emerging: conservation and naturalism.

As noted in some of the discussions in this volume, revolutionary notions were afoot at the turn of the century. In the American zoo community, first at the National Zoo and later at the New York (Bronx) Zoo, William T. Hornaday, championed by Smithsonian secretary Samuel P. Langley and other influential persons of the time, sowed the philosophical seeds of creating naturalistic refuges in urban landscapes where disappearing species could be bred. Hornaday's greatest concern was to initiate conservation measures for the nearly extinct American bison. His efforts and those of others paid off handsomely—"buffalo" are abundant today. It took nearly seventy years for the American zoo community to take to heart his concerns for threatened wildlife, but conservation via the propagation of endangered species is now a major emphasis of virtually every modern zoological park in North America.

In Germany, at about the same time Hornaday was promoting his ideas, Carl Hagenbeck was implementing his. Hagenbeck, in his Tierpark outside Hamburg, created naturalistic "panoramas"—a revolutionary and revolutionizing leap forward in zoo design. Zoos would never look the same again. Hagenbeck's 1907 Polar Panorama exhibited polar bears with seals and other Arctic creatures together in a large enclosure complete with artificial ice flows and snow banks. Hidden barriers, not visible to viewers, separated predators from prey. Hagenbeck's next panorama—African Plains— was equally astonishing.[1]

Mimicking Hagenbeck's accomplishments became an objective of many larger zoos, but it took a long time for zoos to abandon the emphasis on small enclosures and large numbers of different (better yet, rare) species—the "art gallery" approach. Gradually, as the twentieth century progressed and displays became old and outdated, zoos began replacing them with large, naturalistic, multispecies exhibits that encouraged propagation. Increasingly, breeding disappearing species became a significant effort of many zoos. Today, a dozen or more animal species (and the number will continue to grow) owe their survival to zoo propagation programs. Currently, the zoo community of North America has pooled its resources to establish nearly seventy species survival plans for endangered animals. For these species, such programs are very likely their last chance.

This second transition—the integration within zoos of scientific studies, conservation methods, educational displays, and naturalistic exhibits in the latter half of the twentieth century—is an equally exciting story and the material of the sequel to this volume. But it is clear that the underpinnings of the mission of zoos at the end of the twentieth century were set down and spelled out at the end of the nineteenth century.

APPENDIXES

The Value of
Old Photographs
of Zoological
Collections

*T*he collation and study of old photographs is of the greatest importance in an examination of the history of zoos. They show us, more precisely than drawings or verbal descriptions, what the zoos of the past looked like and how their inhabitants were housed and managed. More generally, they may confirm what we already know or strongly suspect. They also evoke the atmosphere of these institutions in the past and furnish a clue to the function they played in society. A passage from Heini Hediger's book *Man and Animal in the Zoo* (1969) explains very well the significance of records, of which photographs form an important part:

> [T]he lack of respect shown to old buildings, and the summary way in which outmoded structures are dealt with is greatly to be deplored. When a decision has been made to build a new animal house . . . the zoo director is so delighted that he arranges for the demolition to go ahead without taking the trouble to record, even in pictorial form, what the old building with all its ancient equipment [looked] like.
>
> This attitude is understandable but it leads to the irretrievable loss of historical material as, unfortunately, there is still not a single . . . museum [for zoos] anywhere in the world today. . . . [S]ome of the earlier breeding successes are no longer achieved today and it is possible that we [could] learn more about the reasons for this by studying the external conditions under which the animals [once] were kept.[1]

Despite the decades that have passed since Hediger's plea, no zoo museum has been established, and this failure gives photographic records an enhanced value.

Moving from emphasis on zoo buildings to their inhabitants, we must remember that until the last few decades, almost all the specimens exhibited in zoological collections were either wild caught or the immediate offspring of wild-caught animals. This is no longer so, and as time passes, stock of any one species is likely to become hybridized and inbred; eventually the animals may lose some of their resemblance to their wild ancestors. In these circumstances, the importance of photographs is obvious and need not be labored. An excellent example of the use of photographs in this context is Dr. Erna Mohr's 1959 work on Przewalski's wild horse.[2] Since her time, many studbooks (pedigree files) contain photographs of key breeding animals.

We can also learn about the way in which animals were handled. Thus the photograph of the London Zoo's Asiatic bull elephant, Dr. Jim, reveals the unnecessary risks taken by Edwardian era keepers. There are few modern primate keepers who can suppress a shudder when they are shown photographs of children embracing a subadult male chimpanzee (capable of ripping them limb

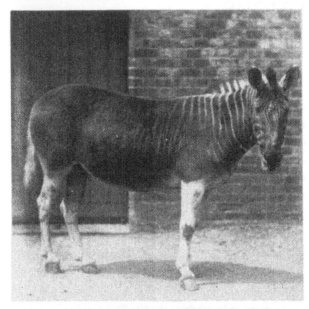

Quagga (*Equus quagga*): one of two photographs taken at the London Zoo by Frederick York, summer 1870. Three additional photos of the same animal, probably taken by Frank Haes in 1864, still exist. This female, purchased from animal dealer Carl Jamrach in 1851, was the only one of her kind to be photographed alive. Her skin is now imaginatively exhibited in the Royal Scottish Museum, Edinburgh, Scotland, and her skeleton is in the Peabody Museum at Yale University. The last quagga died in Amsterdam in 1883. (Collection of John Edwards)

Syrian wild ass (*Equus hemionus hemippus*) at the London Zoo, photographed by Frederick York about 1870. This subspecies, the smallest of the modern equids, is believed to have become extinct about 1930. (Collection of John Edwards)

Sumatran rhinoceros (*Dicerorhinus sumatrensis*) at the London Zoo, August or September 1872. A relatively common exhibit in late-nineteenth- and twentieth-century zoos, this, the smallest of living rhinos, is today one of the rarest and most sought-after species. (Collection of John Edwards)

Indian rhinoceros (*Rhinoceros unicornis*), "Miss Bet," the London Zoo's second rhino, photographed by Frank Haes in summer 1864. This and a similar photograph taken at the same time, now preserved in the archives of the Zoological Society of London, are probably the first photographs ever taken of a living rhinoceros. (Collection of John Edwards)

Javan rhinoceros (*Rhinoceros sondaicus*) in an Indian zoo (ca. 1900). This species is one of the rarest placental mammals and has not been seen in captivity since one died in the Adelaide Zoo, Australia, in 1907. (Collection of John Edwards)

The Value of Old Photographs | 143

from limb) in the Children's Zoo at the London Zoo in 1938. Other photographs reveal how hoofed stock were often kept on insufficiently hard surfaces, resulting in grotesquely overgrown hooves. On the positive side, other photos provide evidence of the affection that existed between keeper and kept.

An interesting use of photographs is to be found in a 1940 issue of the American periodical *Parks and Recreation*. The superintendent of Sydney's Taronga Zoo used photographs of the aged cow elephant "Jessie," taken during the previous fifty-seven years (the first in 1882, the second in 1939), in order to solicit estimates of her age.[3]

Cape mountain zebra (*Equus zebra*) at the Berlin Zoo (ca. 1905). Despite being known as the "common zebra," this, the smallest of living zebras, was accorded special protection as early as 1656, and by the beginning of the twentieth century it was one of the most prestigious and valuable exhibits in such Western zoos as New York, Paris, Amsterdam, Berlin, and London. At the time of the outbreak of the Second World War there were barely fifty individuals left in South Africa. Since then numbers have increased slowly, but there are none to be seen outside Africa. (Collection of John Edwards)

Thylacines, or Tasmanian wolves (*Thylacinus cynocephalus*), at Beaumaris, Hobart, Tasmania (ca. 1910). This carnivorous marsupial is probably extinct; none has been seen in captivity since 1936. (Collection of John Edwards)

Most important, however, photographs are the best method for showing what extinct animals looked like. Mounted specimens, however well prepared, generally distort and fade with time. Thus we owe our most satisfactory images of many now extinct animals, such as the quagga, the Syrian wild ass, the thylacine, the Bubal hartebeest, Schomburgk's deer, and the pink-headed duck, to the photograph.

Having acknowledged this, we must ask why nineteenth-century zoos made so very little use of photography. The first photographs ever taken in a zoo were almost certainly those taken by the count of Montizon at the London Zoo in the

Caspian tiger (*Panthera tigris virgata*) at the Berlin Zoo (ca. 1895). This sub-species has seldom been seen in zoos and is now probably extinct. (Collection of John Edwards)

Specimens from the Hagen-beck importation of Prze-walski's wild horse (*Equus przewalskii*) at the London Zoo in the spring of 1902. (Collection of John Edwards)

summer of 1852. The photos were exhibited at the Royal Society of Arts in December of that year where they were admired by Queen Victoria. Montizon's reason for working at the London Zoo seems to have been to display his skill as a photographer, rather than to record the appearance of animals. Remember that the long exposures and wet plates of the period meant that many hours could be spent in procuring an image of which a modern child would be ashamed. The Zoological Society of London showed polite interest in one of the images (a

Antelope House at the Dresden Zoo (ca. 1900). It was built in the once fashionable Moorish style for antelope, camels, zebras, and giraffes, and its bars were placed so close together that feeding by the public was impossible—an unusual feature for the time. (Collection of John Edwards)

Asiatic bull elephant "Dr. Jim" at the London Zoo (ca. 1905). The keeper is a long way from the elephant, perhaps the most dangerous zoo animal, despite the fact that it is laden with passengers, including a boy astride its neck. Such overconfidence was almost disastrous, for in 1908 this elephant tried to kill his keeper, who was saved only by the courageous intervention of the head elephant keeper, Charles Eyles. Shortly after this incident, Dr. Jim was sold to the Buenos Aires Zoo. (Collection of John Edwards)

African elephants, "Jumbo" and "Alice," photographed by Frederick York at the London Zoo about 1870. This photograph disproves the frequent assertion that these famous elephants never saw each other. (Collection of John Edwards)

Cape lion (*Panthera leo melanochaitus*) by an unknown photographer in the Jardin des Plantes, Paris (ca. 1860). This subspecies became extinct about 1865. This is the only known photograph of a living specimen. (Collection of John Edwards)

Burchell's zebra, or "bonte-quagga" (*Equus burchelli burchelli*), in the Philadelphia Zoo (ca. 1875). This subspecies, which had no stripes on its legs, is believed to have become extinct about 1915.

The Value of Old Photographs | 147

photograph of a pike) but continued throughout the nineteenth century to lavish considerable sums on illustrating its "Proceedings" and "Transactions" with commissioned watercolors of its most interesting specimens from artists such as Edward Lear, John Gould, or Joseph Wolf. One reason for this may have been that a watercolorist, unlike a photographer, could record the plumage of birds accurately.

Later photographers, such as Frank Haes, Frederick York, T. J. Dixon, George Washington Wilson, or Gambier Bolton, were professionals whose motive was profit, not the establishment of a scientific record. This lack of interest is all the more surprising in that photographs were occasionally used in a scientific context; a celebrated instance of this occurred during a meeting of the Zoological Society of London in 1873 when the secretary, Dr. Sclater, exhibited a photograph of the first pygmy hippopotamus to be brought to Europe. Unfortunately, this print has not survived, which is perhaps evidence that it was regarded as an ephemeral item, too insignificant for preservation in the society's archives. In a similar spirit, the Zoological Society of London continued to illustrate its guide

Javan tigers (*Panthera tigris sondaica*) in the Berlin Zoo (ca. 1908). This, the smallest living subspecies, is now either extinct or reduced to a handful of animals. (Collection of John Edwards)

Pink-headed ducks (*Rhodonessa carophyllacea*) from northeastern India, photographed by David Seth-Smith about 1925, in the collection of Sir Alfred Ezra, council member of the Zoological Society of London, at Foxwarren Park, Surrey, England. The last representatives of this beautiful species, possibly these animals depicted, died in this collection during the Second World War. (Zoological Society of London)

with crude woodblocks until the edition of 1904, despite the fact that almost twenty years earlier the first half-tone illustrations to be published in England were made from photographs taken at the London Zoo. It is also fair to record that in the 1920s, motion pictures were made of certain rare animals in the zoo (including the last thylacine to leave Tasmania) on the orders of Sir Peter Chalmers Mitchell, secretary to the Zoological Society (1903–35).

If the Zoological Society of London did not take photography very seriously in the nineteenth and early twentieth centuries, at least it did not actively discourage it, as was the case in other zoological collections of the period. This is demonstrated in almost every chapter of the 1903 classic volume *The Zoological Gardens of Europe,* in which the author, C. V. A. Peel, recorded various methods used to prevent him from using his camera.[4]

On the other side of the Atlantic, the New York Zoological Society prohibited photography in the Bronx Zoo for many years, although, to its eternal credit, it has maintained its own comprehensive and technically superb photographic record of its animals. It is this refusal by so many zoological collections to tolerate

Wallich's deer, or "shou" (*Cervus elaphus wallchi*), in velvet (ca. 1925). Photographed by F. Martin Duncan (1873–1961), librarian to the Zoological Society of London, which now owns his collection of photographs. This specimen, which died in 1926, was presented to the society by its patron, King George V, in 1912 and was the only specimen ever to reach a Western zoo. Note the very distinctive long, sinuous, "pixie" ears, unlike those of any other living deer. By the early 1980s it was believed to be extinct, but in 1987 specimens were discovered in a menagerie of native animals in the gardens of the former Dalai Lama's palace in Lhasa, Tibet. (Zoological Society of London)

photography within their confines, and not bias on the part of the present writer, which explains why so many of the reproductions that accompany this appendix were taken at the London Zoo. Even there, however, many of the most significant early photographs were taken as a result of private enterprise and with little or no idea of their significance. Such as have survived the passage of time are now to be found either in private collections or, as in the case of the Zoological Society of London, in former private collections that have subsequently come to rest in its archives.

If part of the motive for studying our photographic records is to improve wild animal husbandry, it must not be forgotten that this is not an end in itself. Our reason for wishing to maintain animals is that civilized people take pleasure in contemplating the creatures with which they share the planet. It would be a sad day if our posterity had only photographs to explain the source of earlier generations' delight. Viewed in this way, photographs become silent warning of awesome eloquence and assume a value beyond that of mere records.

THE ARCHITECTURE
OF THE NATIONAL
ZOOLOGICAL PARK

*T*he National Zoological Park in its conception differed substantially from preceding zoological gardens. Forged out of a concern for the decimation of the American bison, the park was distinct in its scientific aspirations for a wildlife preserve. Equally significant in the early definition of the park concept was the picturesque aesthetic espoused by Samuel Pierpont Langley, the Smithsonian secretary overseeing the project.

Arguing for the establishment of a national zoological park, Langley wrote to Congress extolling the picturesque qualities of the Rock Creek valley, located just outside Washington's then urban area.[1] In contrast with the small urban zoological gardens maintained in Cincinnati, Philadelphia, New York, or any of the European capitals, Langley lobbied for an expansive scientific station. The Rock Creek site played a great role in the formation of this concept. The untamed, natural beauty of the region inspired in Langley a romantic vision of a haven not only for threatened and endangered native animals but also for the people. In writing to Congress, the secretary prized the varied nature of the land, hailing it as a place where all manner of animals and birds could find an environment similar to their native habitat. He drew attention to the fortunate preservation of the area's picturesque character, despite encroaching urban development, and the importance of preserving it in perpetuity. He also demonstrated that the proximity of "this singularly interesting spot" to the city made it accessible to both the rich and the poor.

In articulating the "picturesqueness of the locality," Langley revealed that the tenets of eighteenth-century English romantic landscape tradition infused his vision for the park. English gardens, exemplified by Henry Hoare's Stourhead in Wiltshire (ca. 1745), gave definition to the notion of "picturesque" by exalting the qualities of roughness, irregularity, and variety present in the wilderness. The cultivation of English gardens was calculated to give the appearance of wild, untouched nature.[2] In the Rock Creek valley, the picturesqueness of the locale was abundantly evident. Langley's concern lay with the conservation of this land's beauty, more than with its cultivation. Combined with this concern, his efforts to secure the property for threatened North American species and to provide accessibility for local people made this park a distinctly American venture and one that Langley saw as properly falling within the Smithsonian Institution's mandate for the "increase and diffusion of knowledge."

The idea of a wildlife park that maintained the natural environment was tied to the development of the national parks in the West. During this time there was growing public consciousness that natural resources were finite; the wilderness and the great game animals could be exploited to the point of exhaustion. As the age of the frontier came to a close, the great abundance and the sense of vast space which had permeated the American psyche seemed in danger. Propelling the conservation of these lands was a recognition of the value of human contact with the natural world.[3] In the eighteenth century, the English, faced with a longing to recover raw nature, had re-created the "wild" within the confines of their private pleasure gardens. In nineteenth-century America, Langley's zoological park-cum–picturesque retreat took its cue from the country's great urban parks, such as Central Park in New York, which were democratic institutions, accessible to all.[4] Of a magnitude akin to the federal measures preserving the majestic spaces of the West, with all their flora and fauna, the National Zoological Park's true distinction lay in its scale and scientific aspirations.

Throughout the years after the National Zoo's establishment, Langley reiterated these points, actively comparing the park with its counterparts both in Europe and in the United States:

> The natural advantages possessed by the National Zoological Park are unrivaled. I am not aware that any city has a zoological park of such extent, so picturesquely situated, and so easily accessible. It is possible here to provide almost any exposure which animals may require, to give them large paddocks with abundant shade and seclusion, or to furnish for waterfowl and other aquatic animals lakes, ponds, and long reaches of water, where they can live in almost perfect natural conditions. These desiderata for the animals may be combined with landscape effects; with stretches of meadow and wooded hills; with overhanging cliffs and running water, with dense shade and sunny slopes; such as can not fail to prove a lasting advantage and pleasure to the public.[5]

Langley's concept of the park was shaped by several circumstances. The original impetus for the preservation of native animals conceived by William Temple Hornaday and the strictures placed on the park's development on account of the funding arrangement devised by Congress both contributed significantly to the formation of the park.

William Temple Hornaday became aware of the precarious condition of the American bison in the West through his work as chief taxidermist of the National Museum. On an expedition to collect bison specimens for the museum in 1886, he was profoundly affected by what he witnessed of the rapid disintegration of these great herds out west. Realizing that greater efforts were needed to prevent the animals' extermination, he argued for "a suitable place in which to preserve representatives of [the] great game animals" before they were all exterminated. "The buffalo is already practically extinct, so far as his wild state is concerned."[6] Upon his return to Washington, he published an account of his trip, *The Extermination of the American Bison,* and he later established the National Bison Society for the preservation of the animal. The park as Hornaday imagined it would have provided a reserve primarily for the breeding of North American animals, with only limited public access.

The efforts to pass the measure in Congress, a battle waged for two consecutive years, 1888 and 1889, also shaped the park's development. Hesitation in the House of Representatives over the scientific merit of maintaining a collection of animals resulted in a motion to have the city of Washington provide half of the zoo's budget. If the citizens of the District were to be the primary beneficiaries of what a number of congressmen viewed as little more than a local menagerie, then support for the park should come from local tax money. The bill as it finally passed dictated that the appropriations be evenly split between the federal government and the District. As has been well argued by Helen Horowitz, this funding arrangement essentially prevented the early concept of the park as wildlife reserve from ever being realized.[7]

Langley, upon assuming the position of secretary in 1887 after Spencer Baird's death, adopted the park as his personal project. In making the Rock Creek area accessible for both animal maintenance and people, Langley sought a style of architecture and planning sensitive to the innate scenic beauty of the site. He looked to Boston for the best architects, soliciting landscape architect Frederick Law Olmsted to design the grounds and William Ralph Emerson to design the buildings.[8] These two men represented the apogee of the "picturesque" at that time in American architecture and landscape design.

It was Olmsted and Emerson whom Langley had in mind when he dictated that no action was to be taken regarding the park, "even to the laying out of a foot path," without the explicit sanction of "a landscaper and architect of approved reputation."[9] Langley further stated that he wished to be informed of every decision, whether to fell a tree or plan a road. His intensive hands-on direction, coupled with the decision to pass control to Assistant Secretary George Brown Goode rather than to Hornaday in the case of the secretary's absence, precipitated Hornaday's resignation before the zoo ever opened to the public. Almost immediately, Langley hired Frank Baker, a Washington, D.C., resident and a professor of anatomy at Georgetown University, to fill Hornaday's place. New to the field of zoo management, Baker was likely to have been more willing to follow Langley's lead.

The most prominent landscape architect in the country at the time, Frederick Law Olmsted was close to the end of his career when Langley wrote him in May 1890. Nonetheless, his acute interest in the National Zoo project was evinced in his immediate acceptance of Langley's invitation; he responded that he was in fact "fairly familiar with the ground in question and, nearly thirty years ago, and often since, [had] urged upon [his] friends in Congress the peculiar advantages of its topography for a public park."[10] Olmsted had demonstrated this subtle and orchestrated enhancement of natural features, most notably in New York's Central Park, one of his earliest projects. In designing his many parks throughout the country, Olmsted worked toward a unified statement of naturalism amid urban development. With Central Park's design, he provided "an antithesis to [New York's] bustling, paved, rectangular, walled-in streets," creating "lungs for the city" while considering separation of traffic, recreation for the people, and the integration of required structures into the landscape.[11]

Although Olmsted's achievements represented what Langley sought for the park, the secretary acknowledged from the very beginning that the Smithsonian could ill afford the Olmsted firm's services. The tenuous system of year-by-year appropriations precluded the possibility of any long-term planning. Nevertheless, the Olmsted Associates, taking over the supervision of the project from the elder Olmsted, worked relentlessly to maintain their role as consultant to the project. The reasons for their dedication are illuminated somewhat in a letter to the senior Olmsted on the importance of maintaining "a professional interest in a work in Washington leading up so closely . . . to the proposed Rock Creek Park."[12]

During the summer of 1890 the Olmsteds worked with Langley to develop a "pleasure drive" to run east-west through the park. The animal houses were to be clustered together in the center of the property. Though this limited development was virtually necessitated by the lack of funds, it nevertheless served Langley's purpose by conserving the majority of the park in its natural state. In arguing for its creation, Langley had hailed the park as an unprecedented opportunity for a zoo unlike any other. The development of this small gardenlike cluster of houses, however, closely mimicked that of other, more urban zoos. What was unique about the National Zoological Park was its sylvan setting, and the vast and unruly beauty of it all Langley sought to showcase to the public: "My guiding wish is to hold to the first and well considered plan of bringing the animals chiefly together in one relatively limited but absolutely large area of thirty or forty acres, and treating the rest of the Park with reference to its natural beauties." [13]

He set aside specifically for this purpose more than twenty acres of the park, which became known as Missouri Valley. For Langley this part of the park (today known as Beaver Valley) represented the height of the picturesque retreat, where "the animals are the accessory to the landscape, and not the landscape an accessory to the animals."[14] The only animals in this area during the park's first decade were beavers brought from Yellowstone Park.[15] The beavers, making their homes, could only heighten the picturesque effects of the valley. Langley's preoccupation with Missouri Valley was such that Frederick Law Olmsted Jr.

referred to it privately as Langley's "pet valley."[16] Indicative of the emphasis that Langley placed on the aesthetic value of a place, the secretary observed, "There should be more picturesque bridges over the little stream at the bottom of 'Missouri' valley."[17]

The image of the rustic, Arcadian landscape, as expressed in Olmsted's work, found its architectural parallels in the buildings of Boston architect William Ralph Emerson. For the previous twenty years, Emerson, in the vanguard of what became categorized by Vincent Scully as the "stick and shingle" style, had been providing grand houses in a uniquely rustic style for wealthy patrons from Bar Harbor to Newport.[18] He was praised for having "taught his generation how to use local materials without apology, but rather with pride in their rough, homespun character."[19] His sprawling houses, with their asymmetrical and open plans, were sophisticated interpretations of the natural environment. Carefully sited on rocky, coastal promontories or in the midst of the wooded suburbs outside Boston, these houses heightened their picturesqueness through the effective and combined use of materials: shingles of a variety of shapes and sizes, often stained to achieve certain hues, timber frame ornament, and unwieldy, rusticated stone. In her monograph on Emerson, Cynthia Zaitzevsky explained that he "showed remarkable sensitivity in adapting his houses not only to topographical peculiarities but to the special mood of their surroundings. In keeping with his feeling for site, Emerson selected the materials of his buildings with an eye for making the houses appear as much as possible a part of nature."[20]

William Ralph Emerson would clearly provide an appropriate architecture, one that would take its cue from the surrounding Rock Creek area. Langley found in the architect a philosophical approach resonant with his own. In 1889 Emerson had published a series of lectures on freehand drawing, stressing reliance on self-taught truths, individualism, and inspiration from nature.[21] These articles were illustrated with five sketches, one of which had been drawn with a toothpick. The expressive, organic quality of Emerson's spontaneous sketches emblemized the design aesthetic that Langley sought for the zoo. The secretary requested his advice for even the smallest of jobs, grateful to receive "merely one of [Emerson's] famous 'toothpick' sketches, which [would] give some adumbration of [the architect's] idea of how such a building should be treated."[22] Emerson's bridges and fences, as well as the animal houses and the aviary, as Langley pictured them, would complement and be carefully sited within a park laid out with Olmsted's guidance.

Even as Langley sought out Emerson for his picturesque aesthetic, the Smithsonian was consulting with other zoos to discuss appropriate zoo architecture. With little experience in animal collecting or park maintenance between them, Langley and Baker were introduced to many practices through meetings with directors from other zoos. As a result of this exchange, the fashionable idea of associationism was incorporated into the idea of the rustic retreat from the city.[23] The architectural style of animal houses took its cue from the typical human architecture of the animal's native region. Buildings following such aesthetic guidelines were especially popular in European zoological gardens during this period. Great Indian temples were constructed for the elephants, such as the one at the Berlin Zoo, laden with fantastical animal sculpture and paintings. These were colorful, opulent palaces for the people rather than the animals, coming at the height of a period of cultural and political imperialism. In the United States, at a park whose primary occupants were to be North American species, it seemed only logical to provide housing reminiscent of the American frontier: the log cabin. Arthur Brown, director of the Philadelphia Zoological Society, articulated this opinion in a visit with Langley, Superintendent Frank Baker, and Frederick Law Olmsted. Brown felt that "for distinctively North American animals, a distinctively North American building should be provided in the log cabin, which [was] at once strong, durable and cheap."[24]

Emerson's first house for the zoo, the Buffalo House, was a glorified log cabin. An enormous two-level wooden frame structure, it was a kind of double-apsed basilica, complete with double transepts and clerestory windows. The house was finished with a veneer of large, roughly hewn logs with the bark left on. Emerson used the cumbersome timber that seemed to embody the American West in a fashion similar to his creative manipulation of the shingles and clapboards in his New England houses. The complex plans of his domestic architecture, however, were simplified in the design of the buffalo barn; the building was divided on a cross axis into four enclosures, its entrances defined by large, rounded arches.

The romantic wooden Buffalo House was followed by the first masonry construction at the zoo, the Carnivora House. Designed by Emerson in 1891 to house virtually all the animals until further structures could be erected, the Lion House (as it was later called) was the ideal park building. It was constructed of massive, irregular, rough-hewn stone and featured a grand rounded-arch brick entrance. Two corner towers, clearly separate from the animal enclosures, provided large windows and gentle, curving balconies, inviting visitors to enjoy the views. The use of the Rock Creek gray gneiss stone, while emphasizing the structure's connections to the setting, also created an impression of strength, reminding visitors of the great predators housed within.[25] With its austere facade of cages and two corner towers, which served as picturesque lookouts but iconographically also as watchtowers, the building carried the additional unintended connotation of a prison. As Emerson had done with his majestic houses on the secluded bluffs of Maine, he employed the materials of the region to provide in the Carnivora House a solid place of repose for the visitors.

Although the Carnivora House represented the kind of architecture Langley sought for the park, it also revealed the problems of working with Emerson. Upon receipt of the drawings, Baker wrote to the architect that he was "somewhat appalled at its dimensions."[26] Even beyond the question of financing the scheme was the problem of executing it from Emerson's trademark artistic sketches. Langley was forced to find a local architect to create working drawings; he wrote to Emerson: "I hope to still avail myself continuously of your valuable services, but I find that it will be imperatively necessary to employ some person here, upon the ground, to prepare working drawings and details from the sketches furnished by you."[27] It would seem logical that in 1891, when only half the appropriation asked for was received, these conditions would have precluded further solicitations from someone like William Emerson, distantly located and an unreliable contractor. Langley, however, in his personal quest for an architecture sympathetic to the land, so clearly articulated in Emerson's work, repeatedly overlooked the architect's delays and lack of practicality. In the same letter chastising Emerson for the extravagant scale of the Carnivora House, Baker wrote that the secretary had requested from the architect a "belvidere [sic] to surmount the highest hill in the park." This commission was significant in its reflection of Langley's goals for the zoo. An observation tower could not serve

The Buffalo House, built in 1891, evoked for visitors the log cabins of the American frontier. (Courtesy of the National Zoological Park)

The Carnivora (Lion) House was the first large, permanent animal house built at the zoo. (Courtesy of the Smithsonian Institution)

the animals; it benefited only the visitors, providing yet another venue by which to enjoy the natural beauty of the Rock Creek site.

Emerson's design for the belvedere was no doubt the epitome of what Langley sought for the park. The wooden tower emerged from an elevated foundation of rough-hewn stone. Its clapboard siding protruded in a delicate and highly unusual imitation of log cabin–style construction. Heightening the tower's asymmetry, balconies descended along one side, covered by an ornamental paneled gable. Narrow decorative mullioned windows would have contributed to the romance of a dimly lit climb to the sweeping view of the park achieved at the summit. Although never built, this belvedere evidenced Langley's continued, idiosyncratic quest for the picturesque retreat.

Despite having sought out local architects, Langley was persistent in his efforts to create a unified scenic retreat with Emerson as chief designer. There was, as he often avowed to Emerson, "no one whose professional skill [he valued] more."[28] In this 1894 letter Langley requested designs for an elephant house, a public comfort house, a gate house, and other structures.

The Elephant House, intended specifically for Asian elephants, provided Emerson with an opportunity to explore exotic stylistic associations. His sketch described a long, low-lying structure with paddocks at either end. A pagoda-shaped roof, echoing the slope of the hillside site that Emerson suggested for the house, elicited images of a Japanese temple. The building was relatively unadorned, with the simple latticework frames of the windows echoed in the scalloped brackets supporting the roof and the balconies overlooking the yards. Two elephantine columns announced the entrance, which was a small curved porch that echoed the line of the main roof. The house was to be cut into the bank of the hill below where the Carnivora House sat, carrying "the building sufficiently back from the river to allow an unbroken view up the valley from the bridge."[29] Emerson seemed to have considered the placement of the building in the site more substantially than he did the animals' needs. The design, while praised by Langley and later given to other architects as a stylistic model, was criticized for lacking a lofty room for the animals.[30]

The Smithsonian returned to Emerson at every opportunity for a permanent or substantial building for the park. This was the case a few years later when the zoo needed a building to accommodate a growing variety of animals that required a higher temperature than that maintained in the existing principal animal house. The Antelope House, as it was called within a year of its erection, provided a rustic country house appropriate to the park setting.

Of all Emerson's designs for the park, the Antelope House most closely resembled his New England houses. Its sprawling, L-shaped plan distinguished it from the other sheds and animal houses. In a park whose buildings were heated with steam from a central location, the slender chimney atop the house's hipped roof immediately identified the building as a residence. Large, overhanging gables were decorated with curved framework and supported by wooden scal-

This romantic lookout tower, the belvedere, although never built, skillfully articulated Secretary Langley's ideal for the zoo as a picturesque retreat. (Courtesy of the Smithsonian Institution)

William Ralph Emerson's 1894 design for an elephant house, never built, recalled images of Eastern temples or palaces with its pagoda-shaped roof. (By permission of the Library of Congress)

The Antelope House of 1897, initially designed to house reptiles, monkeys, and tropical birds, recalled Emerson's New England houses. (Courtesy of the National Zoological Park)

loped brackets such as Emerson had employed earlier in the Elephant House design. The most distinctive feature of the house, however, was the clerestory register, marked by crossed timber brace frames. This decorative motif, reminiscent of barn architecture, provided the primary connection to the building's purpose in the park.

In 1899 Langley again contacted Emerson, concerning designs for an aquarium, windmill, and watermill and revisions to the Elephant House design. There was no subsequent mention of the windmill and watermill, and it seems unlikely that these were being explored as new sources of power for the zoo. In all likelihood they were probably Langley's latest romantic follies, efforts to enhance further the appearance of the park. Emerson was obviously in tune with such wishes, writing to inform the secretary of his wish to "devote [his] time to making sketches for the various picturesque buildings . . . [as he did] not wish to lose the opportunity of helping [Langley] in these beautiful and picturesque schemes."[31] However, Superintendent Frank Baker later wrote to Langley, "Although this matter has been fully discussed with Mr. Emerson and he is aware of the urgency of the case, no new plans have been furnished involving the special features desired by you. It seems unlikely that anything can now be done and I suggest that some local architect be employed to prepare plans and go on with the work."[32]

It was inevitable that at some point working with Emerson would become completely unfeasible. Because the restrictive funding of the zoo necessitated for the most part temporary structures (although these buildings often remained for decades), Langley's emphasis on the visual messages of architecture posed particular challenges to Baker and others. It often fell to these subordinates to solicit from the architects a design that would be satisfactory to Langley—to convince them to produce, "by any little subterfuge known to architects, as picturesque an appearance as possible."[33] In Hornblower and Marshall, the local architects who took over the design of buildings at the zoo, the Smithsonian found a firm with a working style resonant with Langley's idiosyncratic vision.

Although Hornblower and Marshall were not exactly proponents of the picturesque, their responsiveness to Langley was what probably obviated the need to return to Emerson for design ideas. At the time they took on work at the park, Hornblower and Marshall were simultaneously becoming heavily involved in numerous Smithsonian projects, including the renovations to the original Smithsonian building and its neighbor on the Mall, the National Museum building (Arts and Industries). They soon assumed the role of institutional architects, responsible for everything from the design of tables and fireplace surrounds to the creation of the monumental Beaux-Arts–style National Museum of Natural History. Two other projects in particular, the Art Room of 1899 and the Children's Room of 1901, both in the original Smithsonian building, evinced the primary role that Secretary Langley played in setting the course of design development and the immense weight that he placed on a project's aesthetic and decorative intent.[34]

In his efforts to enrich the beauty of the park and the experience of the visitor, Secretary Langley requested advice from Hornblower and Marshall on how to execute his scheme for a horse drinking fountain: "It is a polished red granite trough allowing for two horses to drink together, fed by water flowing from a cup held aloft in the hands of a Centaur, the figure of the man and in part the figure of the horse being in very high relief on a heavy granite background."[35] This was not the first of such whimsical requests. Originating with the belvedere commissioned from William Ralph Emerson, it was only one of Langley's many moves to create in Rock Creek Park his ideal retiring grounds. In 1894 he had asked the Olmsted Associates for a formal maze such as might be found at Hampton Court in England. The firm, viewing the proposal as thoroughly inappropriate for the picturesque park, had responded: "While we have found that there is always much to be said in favor of suggestions coming from Secretary Langley . . . yet in this instance we are reluctantly obliged to give our opinion decidedly against his propositions. . . . [S]uch a maze in a public ground of the character of the Zoological Park and in the climate of Washington would be a complete failure."[36]

In working for Secretary Langley it was obviously necessary at times to accommodate such a fertile and fanciful imagination. In fact, it seems that Langley's quirky and difficult nature was common knowledge in the capital; the McMillan Commission, eager to retain the Olmsted Associates for the monumental improvements planned for the Washington park system, requested Frederick Law Jr. to "humor the Secretary" until the Park Commission had progressed further in its planning.[37]

Hornblower and Marshall's most significant project for the National Zoo was the Small Mammal House, the second major animal house at the park. Superintendent Baker wrote to the architects noting that, since the Small Mammal House was to be permanent, it was "of prime importance" to have "a pleasing design, especially for the exterior."[38] This letter elicited from Hornblower and Marshall a scheme dramatically abandoning the rustic vision of the early years at the park. As the architects described it, the building was to be "of brown glazed brick up to the top of the cage doors," and they planned "to continue this brick as pilasters, and to fill the panels between with red bricks, blue-black headers, yellow tones, etc. making a kaleidoscopic effect in brown tones."[39] The roof was to be laid with clay tiles of varying shades of green, and areas of glass were to serve as skylights for the cages within. At nine points on the roof animal finials were perched atop gables. This proposed building provided a striking contrast to any existing structure at the zoo, and particularly to Emerson's Carnivora House, with which it was most readily compared.

The design for the Small Mammal House represented a move away from the rough and irregular texture of the Carnivora House, from its Richardson-inspired architectural massing, from all that exemplified the picturesque. The Roman bricks indicated for the walls were slender and brown glazed, giving a delicate, linear quality to the building. The design was meant to avoid texture; it relied instead on color to convey its decorative intent. Surprisingly, the plans passed through Baker without much comment, and the plans were sent on to Olmsted for advice on the placement of the building in the park, as had become routine.

Because of this tradition of consultation which Langley had initiated, the Olmsted vision can be seen as a principal regulating factor in the formative years of the park. With the Small Mammal House, Olmsted's opinions conserved the zoo's picturesque nature for a little while longer, prolonging the divergence toward other stylistic trends. The Olmsted Brothers rejected the new building's proposed location directly opposite the Carnivora House, preferring instead a place that would make the building "somewhat less strikingly conspicuous, and would injure a less attractive bit of park scenery for the time being."[40] Additionally, they included a private letter of criticism, voicing reservations about the style of the design:

It seems to me in general that it is not desirable to make the buildings of the Zoological Park striking or bizarre. Picturesqueness is perhaps to be desired, but it should be picturesqueness of the unobtrusive and modest kind. . . . I have some fears that the building described in this drawing and the accompanying letter would verge upon the bizarre, with its brown glazed brick and highly interesting exotic tile roof. . . . I do not think it is necessary to go in for exotic forms and materials when very quiet, charming, picturesque effects can be made by a skillful use of the materials and forms which are well acclimated in Washington and fit comfortably into its landscapes.[41]

Hornblower and Marshall's 1903 design for a small mammal house, with its "kaleidoscopic brick" walls, represented a significant departure from the zoo buildings that had preceded it. Olmsted decried the "highly exotic roof" as out of character with the park. (Smithsonian Institution)

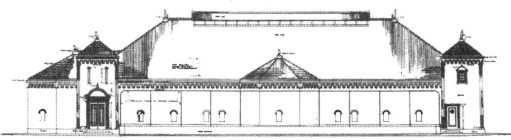

Spurred by these criticisms, Hornblower and Marshall revised their submission with the park's setting in mind. The most significant difference between the two submissions was the change of materials, from the polychrome brick to the gray gneiss of the Rock Creek region. The use of gneiss stone created a building that conformed with the context established by Emerson's Carnivora House and extolled by Olmsted.

The building during its design process had changed radically, such that what had once been labeled "bizarre" was also then called "rustic." Superintendent Baker received a letter inquiring about the Small Mammal House from the commissioner of the Bureau of Fisheries, of the Department of Commerce and Labor: "[T]his bureau, being frequently called upon to plan rustic houses, my attention was drawn through our Engineer, to a very fine building of this character in the Zoological Park just being completed on the hill near the lion house."[42]

It is probable that the building in April 1905 was still lacking a completed roof, for it was the roof, with its array of variegated green tiles and its animated finials, which effected the Small Mammal House's rustic character. The finials, designed by Laura Swing Kemeys, were considered "an integral and necessary part of the structure."[43] Three different model types, lynx, fox, and bear, represented the variety of animals which could be housed within. Hornblower and Marshall, writing to the Perth Amboy Terra Cotta Company, which produced them, instructed that the sculptures have a "strong silhouette," and they advised that the distance from which they would be viewed did not require "any smooth or careful finish."[44]

This original Small Mammal House, the oldest building still in use at the zoo, later became the Monkey House and, later still, the "Think Tank," an

Because of the criticism leveled at their first design, Hornblower and Marshall resubmitted their plans for a small mammal house, using the locally quarried gray gneiss stone that Emerson had used in the Carnivora House. Later converted to the Monkey House, it housed the majority of the zoo's primate collection for decades. After two years of renovation, this structure was reopened in 1995 as the "Think Tank," an exhibit on the nature of animal intelligence. It is the oldest building still in use in the zoo. (Colin Varga/National Zoological Park)

exhibit exploring the nature of animal intelligence. Some ninety years later, the finials atop the roof still denote the house's original function. The exotic roof appeared in 1906, the year of Secretary Langley's death, as the signifier of changing trends at the National Zoo. Although the finished house respected the picturesque aesthetic established by Langley and Olmsted, the building's first design submission in 1903 and the roof, consequently, as the only executed part of that original plan, symbolically closed the period of attempted naturalism and refuge at the zoo.

The National Zoological Park was conceived as an entity distinct from all previous animal collections. Coupled with Hornaday's original dream of a preserve for endangered North American animals was Langley's individual aesthetic for a picturesque retreat. The sylvan setting of Rock Creek inspired this dual vision: the open expanses appeared ideal for grazing and for buffalo ranges; the meadows, in dramatic juxtaposition to the wooded hills and the rocky, winding path of the creek, provided the elements of the picturesque park. The funding arrangement created by Congress precluded the establishment of a wildlife sanctuary. It heightened the urgency of attending to the desires of a local taxpaying public. Although the meager funds prevented the development of the park along a unified plan such as Frederick Law Olmsted had achieved in Central Park, Langley was determined in his creation of the sylvan refuge. This vision was most clearly expressed in his dedicated solicitation of Olmsted and Emerson. Langley's dogged commitment to Emerson throughout the first decade of the park's existence resulted in a handful of architectural gems. With each of these houses, Emerson transformed natural materials and architectural form into a unique expression of the variety, irregularity, and ruggedness of nature's beauty. They remained, most of them well into the twentieth century, monuments to a picturesque ideal.

NOTES

Reflections on Zoo History

1. Lucile H. Brockway, *Science and Colonial Expansion: The Role of the British Royal Botanic Gardens* (New York: Academic Press, 1979). Sally Gregory Kohlstedt, "Museums: Revisiting Sites in the History of the Natural Sciences," *Journal of the History of Biology* 28 (1995): 151–66.

2. See also Harriet Ritvo, *The Animal Estate: The English and Other Creatures in the Victorian Age* (Cambridge: Harvard University Press, 1987).

3. J. Orosz, *Curators and Culture: The Museum Movement in America, 1740–1870* (Tuscaloosa: University of Alabama Press, 1990).

4. Sally Gregory Kohlstedt, "Henry A. Ward: The Merchant Naturalist and American Museum Development," *Journal of the Society for the Bibliography of Natural History* 9 (1980): 647–61.

5. Sally Gregory Kohlstedt, ed., *The Origins of Natural Science in America: The Essays of George Brown Goode* (Washington: Smithsonian Institution Press, 1991). See also Helen Lefkowitz Horowitz's detailed chapter in this volume on the founding of the National Zoological Park.

6. M. J. Lacey, "The Mysteries of Earth-Making Dissolve: A Study of Washington's Intellectual Community and the Origins of Environmentalism in the Late Nineteenth Century" (Ph.D. diss., George Washington University, 1979).

Menageries and Zoos to 1900

1. Gustave Loisel, *Histoire des ménageries de l'antiquité à nos jours*, 3 vols. (Paris: Octave Doin et Fils and Henri Laurens, 1912); George Jennison, *Animals for Show and Pleasure in Ancient Rome* (Manchester, England: Manchester University Press, 1937); Jocelyn M. C. Toynbee, *Animals in Roman Life and Art* (Ithaca, N.Y.: Cornell University Press, 1973).

2. Lord S. Zuckerman, *Great Zoos of the World* (Boulder, Colo.: Westview Press, 1980); Bob Mullan and Gary Marvin, *Zoo Culture* (London: Weidenfeld and Nicolson, 1987); Stephen St. C. Bostock, *Zoos and Animal Rights* (London: Routledge, 1993). See, for example, Bernard Livingston, *Zoo Animals, People, Places* (New York: Arbor House, 1974); Emily Hahn, *Animal Gardens* (New York: Doubleday and Co., 1967); James Fisher, *Zoos of the World*, ed. M. H. Chandler and Vernon Reynolds (London: Aldus Books, 1966); Jon R. Luoma, *The Crowded Ark* (Boston: Houghton Mifflin Co., 1987); Jeremy Cherfas, *Zoo 2000: A Look beyond the Bars* (London: British Broadcasting Corp., 1984); and Hermann Dembeck, *Animals and Men* (Garden City, N.Y.: Natural History Press, 1965).

3. Frederick E. Zeuner, *A History of Domesticated Animals* (New York: Harper and Row, 1963); Hernando Cortés, *Letters from Mexico*, trans. Anthony Pagden (New Haven: Yale University Press, 1986); Bernal Diaz del Castillo, *The Discovery and Conquest of Mexico, 1517–1521*, trans. by A. P. Maudslay (New York: Farrar, Straus and Cudahy, 1956); William Prescott, *History of the Conquest of Mexico*, 3 vols. (Philadelphia, 1860).

4. Fisher, *Zoos*, 25; Jean-Philippe Lauer, *Saqqara: The Royal Cemetery at Memphis* (London: Thames and Hudson, 1976), 19–70, 159, 221. Fisher, *Zoos*, 25; Zeuner, *History*, 417–30.

5. Livingston, *Zoo Animals*, 16; Fisher, *Zoos*, 26; Luoma, *Crowded Ark*, 6. Loisel, *Histoire*, 1:26.

6. Fisher, *Zoos*, 24; Luoma, *Crowded Ark*, 5; Zeuner, *History*, 285–86; A. Leo Oppenheim, *Ancient Mesopotamia: Portrait of a Dead Civilization* (Chicago: University of Chicago Press, 1964), 45–48.

7. Loisel, *Histoire*, 1:37.

8. King James Version, authorized, I Kings 10:19, 20, 22; II Chronicles 9:18, 19, 21, 25; and I Kings 4:26, 28.

9. Loisel, *Histoire*, 1:53.

10. Jennison, *Animals for Show*, 10–27; Fisher, *Zoos*, 28.

11. Jennison, *Animals for Show*, 14–18.

12. Fisher, *Zoos*, 28–29.

13. Jennison, *Animals for Show*, 10–27.

14. Fisher, *Zoos*, 28–30; Loisel, *Histoire*, 1:48.

15. Fisher, *Zoos*, 30–32; Livingston, *Zoo Animals*, 22.

16. Loisel, *Histoire*, 1:31–35; Howard H. Scullard, *The Elephant in the Greek and Roman World* (Ithaca, N.Y.: Cornell University Press, 1974), 132–33.

17. Fisher, *Zoos*, 27; Luoma, *Crowded Ark*, 6.

18. Zeuner, *History*, 463.

19. Fisher, *Zoos*, 27; Luoma, *Crowded Ark*, 6.

20. Loisel, *Histoire*, 1:47–48; C. W. Ceram, *Gods, Graves, and Scholars* (New York: Knopf, 1962), 257.

21. Jennison, *Animals for Show*, 99–125; Toynbee, *Animals in Roman Life*, 16.

22. Jennison, *Animals for Show*, 127.

23. Toynbee, *Animals in Roman Life*, 17.

24. Jennison, *Animals for Show*, 42–62; Toynbee, *Animals in Roman Life*, 17–18.

25. Toynbee, *Animals in Roman Life*, 20.

26. Jennison, *Animals for Show*, 174–76.

27. Ibid., 63–82; Toynbee, *Animals in Roman Life*, 20–22; Loisel, *Histoire*, 1:89–104; Fisher, *Zoos*, 38; Cherfas, *Zoo 2000*, 19; Mullan and Marvin, *Zoo Culture*, 95.

28. Loisel, *Histoire*, 1:103–4.

29. Zuckerman, *Great Zoos*, 5; Cherfas, *Zoo 2000*, 20.

30. Luoma, *Crowded Ark*, 10.

31. Zuckerman, *Great Zoos*, 6.

32. Luoma, *Crowded Ark*, 8–10.

33. Marco Polo, *The Travels of Marco Polo, the Venetian*, edited with an introduction by Manuel Komroff (New York: Horace Liveright, 1930), 106, 145–47, 270.

34. J. J. L. Duyvendak, *China's History of Africa* (London, 1949), cited in Hahn, *Animal Gardens*, 47–48.

35. See Diaz del Castillo, *Discovery and Conquest*, 212–13; Cortés, *Letters from Mexico*, 110–11; Prescott, *History of the Conquest*, 2:116–20; and Henri F. Ellenberger, "The Mental Hospital and the Zoological Garden," in *Animals and Man in Historical Perspective*, ed. Joseph Klaits and Barrie Klaits (New York: Harper and Row, 1974), 62–63.

36. Diaz del Castillo, *Discovery and Conquest*, 212–13; Cortés, *Letters from Mexico*, 110.

37. Diaz del Castillo, *Discovery and Conquest*, 213.

38. Prescott, *History of the Conquest*, 2:116–20.

39. See Hahn, *Animal Gardens*, 41–52, and Fisher, *Zoos*, 40–43.

40. Loisel, *Histoire*, 1:146; Fisher, *Zoos*, 41; on Frederick II as a bird expert, see Erwin Stresemann, *Ornithology: From Aristotle to the Present* (Cambridge: Harvard University Press, 1975), 9–11.

41. Fisher, *Zoos,* 41; Hahn, *Animal Gardens,* 42; Zuckerman, *Great Zoos,* 6.

42. Hahn, *Animal Gardens,* 42; Loisel, *Histoire,* 1:155.

43. Hahn, *Animal Gardens,* 45–46.

44. Mullan and Marvin, *Zoo Culture,* 100.

45. Loisel, *Histoire,* 1:282.

46. Ibid., 1:232; Dembeck, *Animals and Men,* 279–80.

47. Loisel, *Histoire,* 1:247–48; John Barclay Lloyd, *African Animals in Renaissance Literature and Art* (Oxford: Clarendon Press, 1971), 48–49.

48. Loisel, *Histoire,* 1:197–270.

49. Lloyd, *African Animals,* 47; Hahn, *Animal Gardens,* 44; Loisel, *Histoire,* 1:202–3.

50. Lloyd, *African Animals,* 47.

51. Hahn, *Animal Gardens,* 47.

52. Mullan and Marvin, *Zoo Culture,* 98; Hahn, *Animal Gardens,* 45; Loisel, *Histoire,* 1:204.

53. Loisel, *Histoire,* 2:53–69.

54. Fisher, *Zoos,* 50; Luoma, *Crowded Ark,* 12.

55. On the Americas, see Fisher, *Zoos,* 52; on Africa, see Mullan and Marvin, *Zoo Culture,* 103; on elephants in collection, see Dembeck, *Animals and Men,* 283.

56. Fisher, *Zoos,* 50–51; Hahn, *Animal Gardens,* 53–54; Luoma, *Crowded Ark,* 5–12.

57. Cherfas, *Zoo 2000,* 23.

58. Mullan and Marvin, *Zoo Culture,* 103, on its 1752 design; Fisher, *Zoos,* 50–51.

59. Fisher, *Zoos,* 48.

60. Loisel, *Histoire,* 2:104.

61. Mullan and Marvin, *Zoo Culture,* 106.

62. Loisel, *Histoire,* 2:127–64; see also Michael A. Osborne's chapter in this volume.

63. Zuckerman, *Great Zoos,* 8.

64. Ibid., 7–15.

65. Sources for this list are Charles V. A. Peel, *The Zoological Gardens of Europe: Their History and Chief Features* (London: F. E. Robinson and Co., 1903); Zuckerman, *Great Zoos*; and the chapters by Michael A. Osborne, Linden Gillbank, D. K. Mittra, and Vernon N. Kisling Jr. in this volume.

Menageries, Metaphors, and Meanings

1. George Jennison, *Animals for Show and Pleasure in Ancient Rome* (Manchester, England: Manchester University Press, 1937).

2. Gustave Loisel, *Histoire des ménageries de l'antiquité à nos jours,* 3 vols. (Paris: Octave Doin et Fils and Henri Laurens, 1912).

3. Friedrich E. Zeuner, *History of Domesticated Animals* (New York: Harper and Row, 1964).

4. Ibid.

5. Loisel, *Histoire*; see also Barry Lopez, "The Passing Wisdom of Birds," *Orion Nature Quarterly* (autumn 1985).

6. Carl Hagenbeck, *Beasts and Men* (London: Longmans, Green, and Co., 1910).

7. John Dewey, *Reconstruction in Philosophy* (New York: Holt and Co., 1920).

8. Suzanne Langer, *Philosophy in a New Key,* 3d ed. (Cambridge: Harvard University Press, 1957).

9. Emile Durkheim, *The Elementary Forms of Religious Life,* transcribed from French by J. W. Swain (London: Allen and Unwin, 1915).

10. See discussion in Jennison, *Animals for Show.*

11. Wily Ley, *Dawn of Zoology* (Englewood Cliffs, N.J.: Prentice-Hall, 1968).

12. Jennison, *Animals for Show.*

13. Ibid.

14. Ley, *Dawn of Zoology.*

15. Lynn White, "The Historical Roots of Our Ecologic Crisis," *Science* (March 1967).

16. Ley, *Dawn of Zoology.*

17. As quoted in ibid.

18. Ley, *Dawn of Zoology.*

19. Clinton H. Keeling, *Where the Lion Trod: A Study of Forgotten Zoological Gardens* (London: Clam Productions [privately published monograph], 1984).

20. Ernest T. Hamy, "The Royal Menagerie of France and the National Menagerie," in *Smithsonian Institution Annual Report* (Washington, D.C., 1897).

21. Ley, *Dawn of Zoology.*

22. As described by Elizabeth Eisenstein in *The Printing Revolution in Early Modern Europe* (Cambridge: Cambridge University Press, 1983).

23. T. Gill, "Systematic Zoology: Its Progress and Purpose," in *Smithsonian Institution Annual Report* (Washington, D.C., 1907).

24. Eisenstein, *Printing Revolution.*

25. Joseph Klaits and Barrie Klaits, eds., *Animals and Man in Historical Perspective* (New York: Harper and Row, 1974).

26. Gustave Loisel, "The Zoological Gardens and Establishments of Great Britain, Belgium, and the Netherlands," in *Smithsonian Institution Annual Report* (Washington, D.C., 1907).

27. Robert Fitzsimons, *Barnum on London* (London: Bles, 1969).

Zoos in the Family

I wish to thank Clinton A. Fields, Anita Guerrini, Robert Hoage, Herman Reichenbach, and Marvin L. Jones for their assistance in the preparation of this chapter.

1. Gustave Loisel, "Notes sur les ménageries actuelles," in *Histoire des ménageries de l'antiquité à nos jours,* 3 vols. (Paris: Octave Doin et Fils and Henri Laurens, 1912).

2. Lucile H. Brockway, *Science and Colonial Expansion: The Role of the British Royal Botanic Gardens* (New York: Academic Press, 1979).

3. Michael A. Osborne, "The *Société zoologique d'acclimatation* and the New French Empire" (Ph.D. diss., University of Wisconsin–Madison, 1987).

4. Ernest T. Hamy, "Les derniers jours du jardin du roi et la formation du Muséum d'histoire naturelle," in *Centenaire de la foundation du Muséum national d'histoire naturelle,* by the Muséum National d'Histoire Naturelle (Paris: Imprimerie Nationale, 1893).

5. Convention Nationale, *Décret sur le Jardin national des plantes, le Cabinet d'histoire naturelle de Paris, du 10 juin 1793, l'an deuxième de la République: Précédé du Rapport du Citoyen Lakanal, Député de larrège à la Convention, Membre du Comité d'instruction publique* (Paris: De l'Imprimerie Nationale, 1793).

6. Toby A. Appel, *The Cuvier-Geoffroy Debate: French Biology in the Decades before Darwin* (New York: Oxford University Press, 1987).

7. Guy Barthélemy, *Les jardiniers du roy: Petite histoire du jardin des plantes de Paris* (Paris: Librairie du Muséum, 1979).

8. Loisel, *Histoire,* 2:125–30, 3:105–7.

9. Keith Thomas, *Man and the Natural World: A History of the Modern Sensibility* (New York: Pantheon Books, 1983); Harriet Ritvo, *The Animal Estate: The English and Other Creatures in the Victorian Age* (Cambridge: Harvard University Press, 1987).

10. Loisel, *Histoire,* 3:138–41; Archives Nationales, F17 3980.

11. Appel, *Cuvier-Geoffroy Debate.*

12. Ibid.; Goulven Laurent, *Paléontologie et évolution en France de 1800 à 1860: Une histoire des idées de Cuvier et Lamarck à Darwin* (Paris: Editions du Comité des Travaux Historiques et Scientifiques, 1987).

13. Appel, *Cuvier-Geoffroy Debate.*

14. Franck Bourdier, "Isidore Geoffroy Saint-Hilaire," in *Dictionary of Scientific Biography*, 5:358–60.

15. Isidore Geoffroy Saint-Hilaire, "Ménagerie," in *L'Encyclopédie moderne* (Paris, 1823–32), 16:59–64.

16. Isidore Geoffroy Saint-Hilaire, *Note sur la ménagerie, et sur l'utilité d'une succursale ou annexe aux environs de Paris* (Paris, 1860).

17. Paul L. Farber, *The Emergence of Ornithology as a Scientific Discipline, 1760–1850* (Dordrecht, Holland: Reidel, 1982).

18. Michael A. Osborne, "Applied Natural History and Utilitarian Ideals: 'Jacobin Science' at the Muséum d'Histoire Naturelle, 1789–1870," in *Re-creating Authority in Revolutionary France, 1789–1900,* ed. Bryant T. Ragan Jr. and Elizabeth A. Williams (New Brunswick, N.J.: Rutgers University Press, 1992), 125–43; Camille Limoges, "The Development of the Muséum d'Histoire Naturelle of Paris, c. 1800-1914," in *The Organization of Science and Technology in France, 1808–1914,* ed. Robert Fox and George Weisz (Cambridge: Cambridge University Press, 1980), 212–40.

19. Geoffroy Saint-Hilaire, *Note sur la ménagerie.*

20. Isidore Geoffroy Saint-Hilaire, *Acclimatation et domestication des animaux utiles,* 4th ed. (Paris, 1861).

21. Charles V. A. Peel, *The Zoological Gardens of Europe: Their History and Chief Features* (London: F. E. Robinson and Co., 1903).

22. Jérôme Napoléon, "Faits divers," *Bulletin de la Société impériale zoologique d'acclimatation* 6 (1859): 228–32.

23. Michael A. Osborne, *Nature, the Exotic, and the Science of French Colonialism* (Bloomington: Indiana University Press, 1994).

24. Isidore Geoffroy Saint-Hilaire to C. L. Bonaparte, May 4, 1855, MS 2602, Bibliothèque Centrale du Muséum National d'Histoire Naturelle.

25. Pierre-Amédée Pichot, *Le jardin d'acclimatation illustré* (Paris, 1873).

26. Arthur Porte, undated newspaper clipping notes stockholder meeting of February 6, 1903, from *La vie financière,* Archives Nationales, 65 AQ R3157–58.

27. Ibid.; Porte was Albert Geoffroy Saint-Hilaire's successor as Jardin director.

28. *L'Information,* May 6, 1909.

29. Osborne, *Nature,* 127–28.

30. Marc Ambroise-Rendu, "Le Jardin d'acclimatation," *Le Monde,* August 10, 1994.

31. Paul Boulineau, *Les jardins animés: Etude technique et documentaire des parcs zoologiques* (Limoges: Edmond Desvilles, 1934); Jean-Pierre Bechter, "Le parc zoologique du Bois de Vincennes à 50 ans," *12ème Union* 65 (April 1984).

32. Georges Hardy, *Portrait de Lyautey* (Mayenne: Bloud and Gay, 1949).

33. Thomas G. August, *The Selling of the Empire: British and French Imperialist Propaganda, 1890–1940* (Westport, Conn.: Greenwood Press, 1985).

The Order of Nature

1. E. T. Bennett, *The Tower Menagerie: Comprising the Natural History of the Animals Contained in That Establishment, with Anecdotes of Their Characters and Histories* (London, 1829).

2. Thomas Bewick, *The Watercolours and Drawings of Thomas Bewick and His Workshop Apprentices,* ed. Iain Bain (London: Gordon Fraser, 1981); Richard Altick, *The Shows of*

London (Cambridge: Harvard University Press, Belknap Press, 1978); Harriet Ritvo, *The Animal Estate: The English and Other Creatures in the Victorian Age* (Cambridge: Harvard University Press, 1987); Gustave Loisel, *Histoire des ménageries de l'antiquité à nos jours,* 3 vols. (Paris: Octave Doin et Fils and Henri Laurens, 1912).

3. E. W. Brayley, "Some Account of the Life and Writings . . . of the Late Sir Thomas Stamford Raffles," *Zoological Journal* 11 (1827): 382–406; Sir Stamford Raffles, *Memoir of the Life and Public Services of Sir Thomas Stamford Raffles . . . by His Widow* (London, 1830).

4. John Bastin, "The First Prospectus of the Zoological Society of London: A New Light on the Society's Origins," *Journal of the Society for the Bibliography of Natural History* 5 (1970): 385.

5. For general information on the founding of the Regent's Park Zoo, see Adrian Desmond, "The Making of Institutional Zoology in London, 1822–1836," *History of Science* 23 (1985): 153–85, 223–50; P. Chalmers Mitchell, *Centenary History of the Zoological Society of London* (London: Zoological Society of London, 1929); and Wilfrid Blunt, *The Ark in the Park: The Zoo in the Nineteenth Century* (London: Hamish Hamilton and the Tryon Gallery, 1976).

6. Ritvo, *Animal Estate.*

7. Desmond, "Making of Institutional Zoology."

8. Ritvo, *Animal Estate.*

9. Zoological Society of London, Minutes of Council, February 21, 1838, and October 20, 1841.

10. Michel Foucault, *The Order of Things: An Archaeology of the Human Sciences* (New York: Vintage Books, 1970); Yi-Fu Tuan, *Dominance and Affection: The Making of Pets* (New Haven: Yale University Press, 1984); John Berger, "Why Look at Animals?" in *About Looking* (New York: Pantheon Books, 1980).

11. "The Zoological Gardens," *Quarterly Review* 98 (1855): 220, 245.

12. *Report of the Council of the Zoological Society of London,* 1855, 14.

13. *Times* (London), August 12, 1875; *Illustrated London News* 35 (1859): 427.

14. Ritvo, *Animal Estate.*

15. *Report of the Council of the Zoological Society of London,* 1843, 11, 1844, 11.

16. *Visitor's Hand Book to the Liverpool Zoological Gardens* (Liverpool, 1841), 10–14; Bostock and Wombwell's Royal Menagerie, *Illustrated Catalogue* (n.d.; ca. late nineteenth century).

17. Terence Templeton, "The Wild Beasts' Banquet," *New Monthly Magazine* 11 (1824): 362–64.

18. H. Frost, *Biographical Sketch of I. A. Van Amburgh* (New York: Samuel Booth, n.d.), 15.

19. Altick, *Shows of London.*

20. Frank Buckland, *Curiosities of Natural History,* 3d ser. (London: Macmillan, 1900), 109; Frank E. Bedard, *Natural History in Zoological Gardens* (London: Archibald Constable, 1905), 24–25.

21. "Zoological Gardens," 245.

A Tale of Two Zoos

The managing directors of Tierpark Carl Hagenbeck GmbH. generously gave me permission to make and publish reproductions of historical photographs in the Hagenbeck collection; archivist Dieter Beefeld showed me where to look.

1. Gustave Loisel, *Histoire des ménageries de l'antiquité à nos jours,* 3 vols. (Paris: Octave Doin et Fils and Henri Laurens, 1912), 3:125 ff., 291 ff.

2. Charles V. A. Peel, *The Zoological Gardens of Europe: Their History and Chief Features* (London: F. E. Robinson and Co., 1903), 235.

3. Lothar Schlawe, "Aus der Geschichte des Hamburger Tiergartens," *Der zoologische Garten* (Leipzig) 41 (1972): 168, 185.

4. G. Grimpe, "Julius Vosseler zum 70. Geburtstag," *Der zoologische Garten* (Leipzig) 4 (1931): 316.

5. Gustave Loisel, *Rapport sur une mission scientifique dans les jardins et établissements zoologiques publics et privés d'Allemagne, de l'Autriche-Hongrie, de la Suisse et du Danemark* (Paris: Nouvelles Archives des Missions Scientifiques . . . sous les Auspices du Ministère de l'Instruction Publique et des Beaux-Arts, 1907), tome 15, fasc. 3, p. 40.

6. Hans Bungartz, "Zum 65jährigen Jubiläum: Aus der Geschichte unseres Gartens," *Hamburger Zoo-Zeitung* 1, no. 5 (1928): 4.

7. Siegfried Schmitz, *Tiervater Brehm* (München: Harnack, 1984), 169 ff.

8. Harro Strehlow, "Alfred Edmund Brehm als Tiergärtner," *Sitzungsberichte der Gesellschaft naturforschender Freunde zu Berlin,* new ser., 27 (1987): 67 ff.

9. Schlawe, "Aus der Geschichte des Hamburger Tiergartens," 173, 184; annual report, *Zweiter Bericht des Verwaltungsrathes der Zoologischen Gesellschaft in Hamburg an seine Actionäre* (Hamburg: Zoologische Gesellschaft in Hamburg, 1864).

10. Werner Kourist, *Aus dem Tierbestand des Zoologischen Gartens Hamburg* (Berlin: Heenemann, 1969), 4; Schlawe, "Aus der Geschichte des Hamburger Tiergartens," 185; Strehlow, "Alfred Edmund Brehm als Tiergärtner," 69.

11. Werner Kourist, "Die ersten zweihörnigen Nashörner, die Tapire und Wale in den zoologischen Gärten und anderen Tiersammlungen," *Zoologische Beiträge* (Berlin) 19 (1973): 145.

12. Kourist, *Aus dem Tierbestand des Zoologischen Gartens Hamburg,* 6 ff.; Schlawe, "Aus der Geschichte des Hamburger Tiergartens," 173.

13. Kourist, "Die ersten zweihörnigen Nashörner," 141, 143.

14. Schlawe, "Aus der Geschichte des Hamburger Tiergartens," 175; Peel, *Zoological Gardens of Europe,* 97.

15. Schlawe, "Aus der Geschichte des Hamburger Tiergartens," 181 ff.

16. Lee S. Crandall, *The Management of Wild Mammals in Captivity* (Chicago: University of Chicago Press, 1964), 482; Grimpe, "Julius Vosseler," 316; Schlawe, "Aus der Geschichte des Hamburger Tiergartens," 184 ff.

17. Carl Hagenbeck, *Von Tieren und Menschen* (Charlottenburg: Vita, 1908), 31.

18. Heinrich Leutemann, *Lebensbeschreibung des Thierhändlers Carl Hagenbeck* (Hamburg, 1887), 42; Rudolf Meyer, "Ein Gang durch die C. Hagenbeck'sche Handels-Menagerie in Hamburg," *Der zoologische Garten* (Frankfurt) 14 (1873): 25–27.

19. Leutemann, *Lebensbeschreibung des Thierhändlers Carl Hagenbeck,* 16; Theodor Noack, "Neues aus der Tierhandlung von Karl [*sic*] Hagenbeck, sowie aus dem Zoologischen Garten in Hamburg," *Der zoologische Garten* (Frankfurt) 25 (1884–88): 100–115, 326–38; 26:148–49, 170–80, 254; 27:39–47, 75–83; 28:194–202, 273–79, 341–53; 29:59–60, 283–85, 346–48.

20. Herman Reichenbach, "Carl Hagenbeck's Tierpark and Modern Zoological Gardens," *Journal of the Society for the Bibliography of Natural History* 9 (1980): 574; Lothar Schlawe, "Seltene Pfleglinge aus Dschungarei und Mongolei," *Der Zoologische Garten* (Jena) 56 (1986): 308.

21. Hilke Thode-Arora, *Für fünfzig Pfennig um die Welt: Die Hagenbeckschen Völkerschauen* (Frankfurt: Campus, 1989).

22. Wilhelm Fischer, *Aus dem Leben und Wirken eines interessanten Mannes* (Hamburg, 1896), 38 ff.; Hagenbeck, *Von Tieren und Menschen,* 110 ff.

23. Günter Niemeyer, *Hagenbeck: Geschichte und Geschichten* (Hamburg: Hans Christians, 1972), 144, 254.

24. Hagenbeck, *Von Tieren und Menschen*, 162 ff.

25. Josef Menges, postcard dated May 7, 1907, addressed to Herrn Direktor Dr. Seitz, Zoologischer Garten Frankfurt a/Main (private collection).

26. Guidebook, *Führer durch Carl Hagenbecks Tierpark in Stellingen,* 5th ed. (1911), 36 ff., 45 ff.

27. Thode-Arora, *Für fünfzig Pfennig um die Welt,* 173.

28. Cited in William Bridges, *Gathering of Animals: An Unconventional History of the New York Zoological Society* (New York: Harper and Row, 1974), 382.

29. Lorenz Hagenbeck, *Den Tieren gehört mein Herz* (Hamburg: Hoffmann und Campe, 1955), 129 ff.; Ludwig Zukowsky, "Carl-Lorenz Hagenbeck," *Der zoologische Garten* (Leipzig) 21 (1954): 111-14.

30. Grimpe, "Julius Vosseler," 316-17; Schlawe, "Aus der Geschichte des Hamburger Tiergartens," 185-86.

31. Hans Bungartz, "Etwas über die Umgestaltung des Zoologischen Gartens in einen Vogel- und Volkspark," *Hamburger Zoo-Zeitung* 3 (1930): 132-35; Hans Bungartz, "Der neue Vogelpark," *Hamburger Zoo-Zeitung* 3 (1930): 208-11; Erna Mohr, "Der Vogelpark in Hamburg," *Der zoologische Garten* (Leipzig) 4 (1931): 165-69.

Zoos and Aquariums of Berlin

1. Michael Seiler, "Lennés Wirken auf der Pfaueninsel," in *Peter Joseph Lenné: Volkspark und Arkadien,* ed. Florian von Buttlar (Berlin: Nicolaische Buchhandlung, 1989), 171-88.

2. Ibid.

3. Caesar von der Ahé, "Die Menagerie auf der 'Königlichen Pfaueninsel,'" *Mitteilungen des Vereins für die Geschichte Berlins* 47 (1930): 1-24.

4. Ibid.

5. Werner Kourist, *400 Jahre Zoo* (Bonn: Rheinisches Landesmuseum, 1976), 181, 81-98.

6. Ahé, "Die Menagerie," 1-24; Heinz-Georg Klös, *Von der Menagerie zum Tierparadies* (Berlin: Haude and Spener, 1969), 320, 21-24.

7. Richard Béringuier, *Geschichte des Zoologischen Gartens in Berlin* (Berlin, 1877), 36.

8. Klös, *Von der Menagerie zum Tierparadies,* 25-35.

9. Ibid., 29.

10. Ibid.; Michael Seiler, "Peter Joseph Lennés erster Entwurf für den Berliner Zoo: Ein nicht realisiertes Projekt, eine Pfaueninsel vor die Tore der Stadt zu holen," *Bongo* 3 (1979): 63-74.

11. Klös, *Von der Menagerie zum Tierparadies.*

12. Ibid., 58.

13. Ibid., 62-63.

14. Lothar Schlawe, *Die für die Zeit vom 1 August 1844 bis 31 Mai 1888 nachweisbaren Thiere im zoologischen Garten zu Berlin* (Berlin: Selbstverlag, 1969), 65.

15. See, for example, Martin Hinrich Lichtenstein and E. Winckler, *Die veredelte Hühnerzucht,* 2 vols. (Berlin, 1857-58); Theodor Leisering, "Katalepsie bei einem Wolfe," *Magazin für die gesamte Thierheilkunde* 14 (1848): 223-30; Leisering, "Beobachtungen aus dem Berliner zoologischen Garten," *Mag. ges. Thierheilkunde* 19 (1853): 76-106, 203-49, 350-87; and Leisering, "Beobachtungen aus dem Berliner zoologischen Garten," *Mag. ges. Thierheilkunde* 20 (1854): 308-20.

16. Klös, *Von der Menagerie zum Tierparadies,* 58.

17. Ibid., 61–80.

18. Heinz-Georg Klös, "Tierhaltung im alten Berlin: Teil II," *Bongo* 14 (1988): 105–14.

19. Heinz-Georg Klös and Ursula Klös, "Wilhelm Böckmann zum Gedenken," *Bongo* 12 (1987): 73–96.

20. Lothar Schlawe, "Zur Geschichte der Zoologischen Gärten," in *Das Buch vom Zoo,* ed. Robert Keller and Christian R. Schmidt (Luzern: Bucher, 1978), 17–33.

21. Heinz-Georg Klös and Ursula Klös, eds., *Der Berliner Zoo im Spiegel seiner Bauten* (Berlin: Heeneman Verlag, 1990), 401.

22. Heinz-Georg Klös and Dietmar Jarofke, "100 Jahre Tiermedizin in Zoologischen Gärten," *Bongo* 13 (1987): 185–96.

23. Schlawe, *Die für die Zeit vom 1 August 1844 bis 31 Mai 1888.*

24. Heinz-Georg Klös, "Geheimrat Professor Dr. phil. Dr. med. vet. h. c. Ludwig Heck," *Bongo* 11 (1986): 77–96.

25. Klös and Klös, *Der Berliner Zoo.*

26. Harro Strehlow, "Zur Geschichte des Berliner Aquariums Unter den Linden," *Zoologischer Garten,* new ser., 57 (1987): 26–40.

27. Harro Strehlow, "Beiträge zur Menschenaffenhaltung im Berliner Aquarium Unter den Linden I: Der Gorilla (*Gorilla g. gorilla*) M'PUNGU," *Bongo* 9 (1985): 67–78; Strehlow, "Zur Geschichte des Berliner Aquariums Unter den Linden."

28. Strehlow, "Zur Geschichte des Berliner Aquariums Unter den Linden"; Harro Strehlow, "Alfred Edmund Brehm als Tiergärtner," *Sitzungsberichte der Gesellschaft naturforschender Freunde Berlin,* new ser., 27 (1987): 67–80.

29. Alfred Edmund Brehm, *Das Leben der Vögel* (Glogau, 1861), 708; Brehm, *Gefangene Vögel* (Leipzig, 1872), 628, 827.

30. Strehlow, "Zur Geschichte des Berliner Aquariums Unter den Linden."

31. Ibid.

32. Harro Strehlow, "Beiträge zur Menschenaffenhaltung im Berliner Aquarium Unter den Linden II: Weitere Gorillas (*Gorilla g. gorilla*)," *Bongo* 12 (1987): 105–10; Harro Strehlow, "Beiträge zur Menschenaffenhaltung im Berliner Aquarium Unter den Linden: Teil III: Orang Utans (*Pongo pygmaeus*) und Schimpansen (*Pan troglodytes*)," *Bongo* 14 (1988): 99–104.

33. Strehlow, "Beiträge zur Menschenaffenhaltung im Berliner Aquarium Unter den Linden I."

34. Strehlow, "Beiträge zur Menschenaffenhaltung im Berliner Aquarium Unter den Linden II."

35. Strehlow, "Zur Geschichte des Berliner Aquariums Unter den Linden"; Harro Strehlow, "Aus der Vogelhaltung des Berliner Aquariums Unter den Linden," *Gefiederte Welt* 112 (1988): 18–21.

36. Ewald Graetz, *75 Jahre Triton e.V. 1888* (Berlin: Selbstverlag, 1963), 16; Bernhard Blaszkiewitz, "100 Jahre TRITON e.V. 1888," *Bongo* 15 (1989): 105–14.

37. Klös, *Von der Menagerie zum Tierparadies.*

38. Strehlow, "Zur Geschichte des Berliner Aquariums Unter den Linden."

A Paradox of Purposes

1. Lucile H. Brockway, *Science and Colonial Expansion: The Role of the British Royal Botanic Gardens* (New York: Academic Press, 1979).

2. Linden Gillbank, "The Origins of the Acclimatisation Society of Victoria: Practical Science in the Wake of the Gold Rush," *Historical Records of Australian Science* 6 (1986): 359–74.

3. Geoffrey Serle, *The Golden Age: A History of the Colony of Victoria, 1851–1861* (Melbourne: Melbourne University Press, 1963).

4. Michael A. Osborne, *The Société zoologique d'acclimatation and the New French Empire: The Science and Political Economy of Economic Zoology during the Second Empire* (Ph.D. diss., University of Wisconsin-Madison, 1987).

5. See Michael A. Osborne's chapter in this volume.

6. Frank Buckland, *The Acclimatisation of Animals* (Melbourne, 1861), 7.

7. Edward Wilson, "The Alpaca," *Times* (London), July 17, 1858, 10.

8. Edward Wilson, "The Distribution of Animals," *Times* (London), October 20, 1858, 9.

9. Edward Wilson, "The Distribution of Animals," *Times* (London), October 28, 1858, 9.

10. Edward Wilson, "The Distribution of Animals," *Argus* (Melbourne), February 13, 1860, 5.

11. Edward Wilson, "The Distribution of Animals," *Times* (London), September 22, 1860, 10.

12. "Acclimatisation Society," *Times* (London), May 15, 1866, 14.

13. Melbourne Chamber of Commerce, "Proceedings of the Special Committee Appointed to Consider and Report on the Best Means of Promoting Agriculture and Settling the Waste Lands of the Colony," Melbourne, 1855.

14. George Ledger, *The Alpaca: Its Introduction into Australia, and the Probabilities of Its Acclimatisation There* (Melbourne, 1861); George Ledger, "The Alpaca," *Sydney Morning Herald*, February 2, 1864, 3; Gabriele Gramiccia, *The Life of Charles Ledger (1818–1905): Alpacas and Quinine* (London: Macmillan, 1988), 14.

15. Ferdinand Mueller, "Report on the Botanic Garden," *Victoria. Parliamentary Papers*, 1856–57, paper no. 81a, p. 8.

16. "Legislative Council: Introduction of the Alpaca into Victoria," *Argus*, January 26, 1856, 4; "Report from the Select Committee of the Legislative Council on the Alpaca," *Victoria. Parliamentary Papers*, 1855–56, paper no. D.11a; "The Alpaca," *Argus*, February 23, 1856, 4; March 13, 1856, 4; March 18, 1856, 4; March 20, 1856, 5.

17. "Alpacas," *Victoria. Parliamentary Papers*, 1856–57, paper no. 55, p. 2.

18. "Development," *Argus*, April 1, 1856, 4.

19. *Transactions of the Philosophical Institute of Victoria*, "Proceedings of the Meeting of 15 April 1856," xxv.

20. Thomas Embling, "Capabilities of the Colony of Victoria," *Argus*, April 25, 1856, 5.

21. "The Port Phillips Farmers' Society," *Argus*, June 27, 1856, 5.

22. "Alpaca," *Argus*, March 22, 1856, 5; November 18, 1856, 5.

23. "Report from the Select Committee of the Legislative Assembly upon Livestock Importation," *Victoria. Parliamentary Papers*, 1856–57, paper no. D.15a, p. 2.

24. Ibid., p. 7.

25. Ibid., p. 6.

26. *Victorian Hansard* 2 (1856–57): 693, 737, Legislative Assembly for June 2, 1857, and June 5, 1857.

27. Edward Wilson, "On the Murray River Cod, with particulars of Experiments instituted for introducing this Fish into the River Yarra-Yarra," *Transactions of the Philosophical Institute of Victoria* 2 (1857): 25.

28. Edward Wilson, "On the Introduction of the British Song Bird," *Transactions of the Philosophical Institute of Victoria* 2 (1857): 77–88; Royal Society of Victoria Archives, 1857, Latrobe Library, Melbourne, Box 8–9, MS-116633, miscellaneous correspondence file; Ed-

ward Wilson, "Distribution of Animals," *Times* (London), October 20, 1858, 9.

29. Wilson, "Distribution of Animals," *Times,* October 20, 1858, 9.

30. Ledger, *The Alpaca,* 14; "Alpacas in Australia," *Times* (London), February 16, 1859, 9.

31. Ledger, *The Alpaca,* 15–16; *Argus,* October 13, 1858, 4; *Argus,* February 17, 1859, 7; "The Alpacas Presented to Victoria," *Times* (London), May 16, 1859, 12; "Alpacas," *Victoria. Parliamentary Papers,* 1858–59, paper no. A.5.

32. *Victorian Hansard* 4 (1858–59): 15, Legislative Assembly for October 12, 1858; "The Alpacas," *Argus,* November 25, 1858, 7.

33. Wilson, "Distribution of Animals," *Argus,* February 13, 1860, 5.

34. W. Lockhart Morton, "Suggestions for the Introduction of Animals and Agricultural Seeds into Victoria," *Transactions of the Royal Society of Victoria* 4 (1860): 153–57.

35. Lynette J. Peel, *Rural Industry in the Port Phillip Region, 1835–1880* (Melbourne: Melbourne University Press, 1974), 147–48; "Report from the Select Committee on the Experimental Farm," *Victoria. Parliamentary Papers,* 1861–62, paper no. D57.

36. Zoological Society of Victoria, Minute Book, 1857–58, Latrobe Library, Melbourne.

37. "Zoological Society of Victoria," *Argus,* March 1, 1858, 8.

38. *Victorian Hansard* 3 (1857–58): 174, Legislative Assembly for January 26, 1858.

39. Frederick McCoy, "Address in Acclimatisation Society of Victoria," *First Annual Report* (Melbourne, 1862), 33; Zoological Society of Victoria, Minute Book, 1857–58.

40. "Zoological Society," *Argus,* November 6, 1857, 4.

41. Zoological Society of Victoria, Minute Book, 1857–58; "Zoological Society of Victoria," *Argus,* March 1, 1858, 8.

42. Ferdinand Mueller, "Annual Report of the Government Botanist and the Director of the Botanic Garden," *Victoria. Parliamentary Papers,* vol. 2, 1858–59, paper no. 17, p. 4.

43. Zoological Society of Victoria, Minute Book, 1857–58, p. 8.

44. "The Zoological Society," *Argus,* April 9, 1858, 4.

45. Zoological Society of Victoria, Minute Book, 1857–58, p. 20; McCoy, "Address in Acclimatisation Society of Victoria," 33–34.

46. "Melbourne Zoological Gardens Committee," *Victorian Government Gazette,* August 6, 1858, 1497; August 3, 1860, 1439; October 19, 1860, 1961; March 12, 1861, 531.

47. "Useful and Rare Animals," *Victorian Government Gazette,* July 26, 1859, 1553.

48. Ledger, *The Alpaca,* 19; "Alpacas Presented to Victoria," *Times,* May 16, 1859, 12.

49. Ferdinand Mueller, "Annual Report of the Government Botanist and Director of the Botanical and Zoological Garden," *Victoria. Parliamentary Papers,* 1859–60, paper no. 37, p. 8.

50. Acclimatisation Society of Victoria, Minute Book, 1861–63, Public Record Office, Melbourne.

51. Buckland, *Acclimatisation of Animals,* appendix.

52. McCoy, "Address in Acclimatisation Society of Victoria," 36.

53. Acclimatisation Society of Victoria, *First Annual Report,* Melbourne, 1862.

54. "Lands to be Permanently Reserved, etc.," *Victorian Government Gazette,* March 25, 1862, 529; W. A. Sanderson, "Royal Park," *Victorian Historical Magazine* 14 (1932): 114–15.

55. McCoy, "Address in Acclimatisation Society of Victoria," 42–43.

56. Acclimatisation Society of Victoria, *Third Annual Report,* Melbourne, 1864, 7.

57. Christopher Lever, *They Dined on Eland: The Story of the Acclimatisation Societies* (London: Quiller Press, 1992), 74.

58. Zoological and Acclimatisation Society of Victoria, *First Annual Report,* 1872, 19.

Ram Bramha Sanyal and the Establishment of the Calcutta
Zoological Gardens

1. B. V. Subbarayappa, "Western Science in India up to the End of the Nineteenth Century," in *A Concise History of Science in India,* ed. D. M. Bose, S. N. Sen, and B. V. Subbarayappa (New Delhi: Indian Science Academy, 1971), 530–37.

2. C. L. Schwendler, *The Establishment of a Zoological Garden for the Town of Calcutta,* Proceedings of the Natural History Committee of the Asiatic Society of Bengal, March 1873.

3. *Nature* (London), May 10, 1877, 10.

4. *Branch-Industry and Science,* General Proceedings, vol. 222, December 1875.

5. Calcutta Zoological Gardens, "Management Committee Prospectus on the Establishment of a Zoological Garden in Calcutta," December 20, 1875.

6. Calcutta Zoological Gardens, *List of Animals,* printed by the authority of His Excellency, the Lieutenant-Governor of Bengal, Calcutta, 1878.

7. Calcutta Zoological Gardens, "Management Committee Prospectus on the Establishment of a Zoological Garden."

8. A. O. Hume to C. L. Schwendler, January 7, 1876, Records of the Office of the Secretary for Revenue, Agriculture, and Commerce Department, Government of Bengal; C. E. Buckland to J. C. Parker, January 20, 1876, Records of the Calcutta Zoological Gardens Management Committee; C. E. Buckland to J. C. Parker, letter appointing him director of the Calcutta Zoological Gardens for a six-month period, January 25, 1876, Records of the Calcutta Zoological Gardens Management Committee; J. C. Parker, letter accepting position as director of the Calcutta Zoological Gardens, January 21, 1876, Records of the Calcutta Zoological Gardens Management Committee.

9. C. E. Buckland to the secretary, government of Bengal, on the progress of the plans for a zoological garden in Calcutta, January 27, 1876, Records of the Calcutta Zoological Gardens Management Committee.

10. Letter received from the acting assistant secretary, government of Bengal, stating that the Management Committee must, after July 10, 1876, fund the zoo, its animals, and its employees, 1876, Records of the Calcutta Zoological Gardens Management Committee.

11. N. Majumdar, *The Statesman: An Anthology* (a compilation of newspaper excerpts) (Calcutta: The Statesman, 1975), 71.

12. C. E. Buckland, report submitted to the secretary, government of Bengal, on the progress of the plans for a zoological garden in Calcutta, February 8, 1877, Records of the Calcutta Zoological Gardens Management Committee.

13. J. Anderson, "A List of Specimens in the Animal Collection Provided upon Request from the Lt. Governor of Bengal," Records of the Calcutta Zoological Gardens Management Committee, April 20, 1877.

14. J. Anderson, "Statement on Surplus Animals," Records of the Calcutta Zoological Gardens Management Committee, May 9, 1877.

15. List of animals that were included in the collection of the Zoological Garden, Calcutta, 1890, Records of the Calcutta Zoological Gardens Management Committee.

16. J. Anderson to an army officer of the Royal Engineers, Nagpur, Central Provinces, July 29, 1878, Records of the Calcutta Zoological Gardens Management Committee.

17. H. M. Tobin, letter no. 386 to the secretary of the government of Bengal, Financial Department, March 22, 1877.

18. J. Anderson to Dr. Philip Lutly Sclater, secretary, gardens of the Zoological Society, London, letter expressing a desire for a European superintendent, July 14, 1879, Records of the Calcutta Zoological Gardens Management Committee.

19. Secretary Order Book, order no. 5, April 20, 1878, Records of the Calcutta Zoological Gardens Management Committee.

20. Ram Bramha Sanyal, facsimile of Sanyal's resignation submission, April 17, 1895, pars. 2–3, Records of the Calcutta Zoological Gardens Management Committee.

21. C. E. Buckland, resolution not to accept Sanyal's resignation submission sent to the director of public instruction, government of Bengal, August 15, 1878, Records of the Calcutta Zoological Gardens Management Committee.

22. Government of Bengal Resolution, 1878.

23. Resolution to increase Ram Bramha Sanyal's salary, April 24, 1879, Records of the Calcutta Zoological Gardens Management Committee.

24. Honorable secretary's letter no. 50, May 10, 1879, government of Bengal.

25. Government of Bengal to the Calcutta Zoological Gardens Management Committee, April 15, 1880, no. 1514/60, misc.

26. C. L. Schwendler, letter no. 995 from the Honorable Secretary C. L. Schwendler to the secretary, government of Bengal, March 26, 1880.

27. *Calcutta Gazette,* July–December 1899, appendix and supplement (articles on the Calcutta Zoological Gardens and Superintendent R. B. Sanyal); *Annual Report of the Calcutta Zoological Gardens Management Committee,* notes on the need to produce a handbook on the management of animals in captivity (Calcutta, 1888–89).

28. *Nature* (London), August 4, 1892, 314.

29. Response to the Bombay Corporation's request for the advice of a qualified zoologist, February 8, 1894, Records of the Calcutta Zoological Gardens Management Committee.

30. Record of the Proceedings of the Standing Committee of the Bombay Municipality, 1894–95, vol. 18, pt. 2, pp. 173–82.

31. Surgeon Lieutenant Colonel T. S. Weer, agriculture municipal commissioner, Bombay, letter no. 17285 to the secretary, Calcutta Zoological Gardens Management Committee, forwarding a copy of the Bombay Corporation's Resolution #7561 detailing J. M. Doctor to Calcutta for one year, April 25, 1894.

32. Ram Bramha Sanyal, *Hours with Nature* (Calcutta, 1896; Calcutta: City Book Society, 1907).

33. J. Cleghorn, ed., *Scientific Memoirs by the Medical Officers of the Army of India* (Oxford, 1897), pt. 10, 59–94.

34. Results of snake poison experiments sent to the inspector general, Civil Hospitals of Bengal, letter no. 89, June 29, 1899, Records of the Calcutta Zoological Gardens Management Committee.

35. *Annual Report of the Calcutta Zoological Gardens Management Committee,* notes on the management of European zoological gardens, app. D (Calcutta, 1888–89).

36. Note on Rai Ram Bramha Sanyal Bahadur accepted for associate membership, Proceedings of the Asiatic Society of Bengal, 1900, new ser., p. 41.

37. Discussion of R. B. Sanyal's possible retirement, February 15 and April 12, 1902; discussion of training a successor to the present superintendent, February 13, 1904; discussion of R. B. Sanyal's retirement application after thirty years of service, October 21, 1905; Records of the Calcutta Zoological Gardens Management Committee.

38. Revenue Miscellaneous File, government of Bengal, file I2–3, proceedings B, 38–39, March 1904.

39. Proceedings of the Home Department, August 1905, no. 83, par. 5.

40. Government of India, letter no. 2649, August 8, 1905.

41. Review of the qualifications of the new superintendent, June 13 and 27 and July 3, 1907, Records of the Calcutta Zoological Gardens Management Committee.

42. Government of Bengal, letter no. 466 (Revenue Department) on the extension of R. B. Sanyal's services, January 24, 1908.

43. Sanyal's note complaining of his unhealthy housing and its effect on his health, March 9, 1908, Records of the Calcutta Zoological Gardens Management Committee.

44. *Statesman and Friend of India* (Calcutta), October 15, 1908, 5.

45. Daily Register, 1877–1908, Calcutta Zoological Gardens.

American Showmen and European Dealers

1. Gustave Loisel, *Histoire des ménageries de l'antiquité à nos jours*, 3 vols. (Paris: Octave Doin et Fils and Henri Laurens, 1912), 2:41–42.

2. Richard Altick, *The Shows of London* (Cambridge: Harvard University Press, Belknap Press, 1978), 39, 308–17.

3. *Fairburn's Edition of the Life & Death of the Elephant, at Exeter 'Change; to which is added, an account of the Dissection of this noble animal, with other interesting particulars* (London, [ca. 1826]) mentions Cops as keeper on p.16; Sarah Bailey, "The Menagerie in the Tower," in *Strange Stories from the Tower of London,* ed. Alan Borg (London: British Heritage, 1976), 55–59.

4. J. R. Howe to Epenetus Howe, August 13 and September 14, 1834, Somers (New York) Historical Society; "The Oldest of Showmen," *Croton Falls News,* July 17, 1879.

5. William Howe to Epenetus Howe, July 28, 1833, Somers (New York) Historical Society.

6. Robert McClung and Gale McClung, "Capt. Crowninshield Brings Home an Elephant," *American Neptune* 18 (April 1958): 137–41; R. M. McClung and G. S. McClung, "America's First Elephant," *Nature* 50 (October 1957): 400–404.

7. *Boston Independent Chronicle,* May 24 and June 11, 1804; *Salem* (Massachusetts) *Register,* July 27 and 31, 1816; *Niles Weekly Register,* August 10, 1816, 400; *New York Post,* April 5–May 2, 1817.

8. Agreement dated October 11, 1816, between Hachaliah Baily and George Brunn, private collection.

9. For a useful overview of this period, see Richard W. Flint, "Entrepreneurial and Cultural Aspects of the Early Nineteenth-Century Circus and Menagerie Business," in *Itinerancy in New England and New York,* ed. Peter Benes, Dublin Seminar for New England Folklife (Boston: Boston University, 1986), 131–49. See also Stuart Thayer, *Annals of the American Circus,* vol. 1 (Ann Arbor, Mich.: Rymack Printing Co., 1976), and vol. 3 (Seattle: Dauven and Thayer, 1992).

10. J. R. Howe to Epenetus Howe, August 13 and September 14, 1834.

11. Richard W. Flint, "Rufus Welch: America's Pioneer Circus Showman," *Bandwagon* 14 (September–October 1970): 7–8; "Oldest of Showmen."

12. "Estimates of J. R. & W. Howe & Co. Exhibition Nov 21st 1834," Somers (New York) Historical Society.

13. "Articles of Association of the Zoological Institute made and entered into January 14th 1835 at Somers, N.Y.," Westchester County Historical Society, Valhalla, New York. See also an article appropriately titled "Yankee Enterprise" in the *Springfield* (Massachusetts) *Republican,* February 28, 1835, about the Boston Zoological Association, which was probably the Macomber, Welch & Co. show variously titled the "Grand Boston" or "New England Zoological Exhibition." Both Zebedee Macomber and Rufus Welch signed the 1835 Articles of Association.

14. P. T. Barnum, *Life* (New York, 1855), 220.

15. For example, Samuel Goodrich—Peter Parley—was a reformer of juvenile literature in America, and among his nearly two hundred volumes, six were published between 1834

and 1850 which dealt with natural history. See Ruth Miller Elson, *Guardians of Tradition* (Lincoln: University of Nebraska Press, 1964), 212–16, 251–54, and passim; John Nietz, *The Evolution of American Secondary School Textbooks* (Rutland, Vermont: Tuttle, 1966), 77–91; and William Martin Smallwood, *Natural History and the American Mind* (New York: Columbia University Press, 1941), 351–53, which also notes that *A Familiar Description of Beasts and Birds* (1813) was one of the earliest publications for children which had a serious scientific attitude almost unaccompanied by moralities.

16. In 1837 the menagerie of Purdy, Welch, Macomber & Co. in its newspaper advertisements referred the public to "the Pamphlet dedicated to the explanation of the wonders of the unrivalled collection . . . " (see, for example, the [Cooperstown, New York] *Farmer's Journal,* May 29, 1837). The same year one branch of the Zoological Institute published *A delineated description and history of all the beasts, birds, & reptiles, contained therein. Noel E. Waring, manager* (New York, 1837), and the year before a thirty-two-page booklet was issued in Philadelphia by an animal exhibition. Buffon, whose splendid *Natural History* was widely used and copied after it was published in Paris beginning in 1750, was cited on broadsides in 1797 advertising America's first elephant (see copies located in Massachusetts at the Newburyport Public Library and the Essex Institute and Peabody Museum in Salem; in Rhode Island at the Historical Society and the John Carter Brown Library, Providence; and at the New-York Historical Society).

17. W. C. Coup, *Sawdust and Spangles* (Chicago: Herbert S. Stone, 1901), 19; Louis E. Cooke, "Reminiscences of a Showman," *Newark Evening Star,* August 26, 1915.

18. William M. Mann, *Wild Animals in and out of the Zoo,* Smithsonian Scientific Series, vol. 6 (New York: Smithsonian Institution Series, [1943]), 4–5, 126, 202. On nineteenth-century zoos in America, see the chapter by Vernon N. Kisling Jr. in this volume.

19. "The Reiche Brothers," *New York Clipper,* April 25, 1885, 88. See also the obituary of Henry Reiche in the *New York Times,* June 17, 1887, and the *New York Tribune,* June 16, 1887, as well as "Recollections of a Showman" in the *New York Times,* June 18, 1887.

20. David Jamieson, "The Wombwells and the Bostocks," *King Pole* (September 1977): 8, quoting the *Scotsman,* April 10, 1872; Hyatt Frost to W. W. Thomas, June 6, 1872, collection of Fred D. Pfening Jr., Columbus, Ohio. The first black African rhinoceros brought to Europe since Roman times was imported by Carl Hagenbeck and arrived at the London Zoo in September 1868. However, in the summer of 1868 Van Amburgh's menagerie, owned and managed by Hyatt Frost, advertised the only rhinoceros in America and occasionally described it as a black rhinoceros or even an impossible "black Asiatic Rhinoceros" (see "It is a fact . . . 11th" on the reverse side of a handbill dated October 13, 1870, for an appearance in Fostoria [Ohio] and headed on the obverse "This is emphatically the most colossal exhibition . . . Van Amburgh & Co's . . . Rhinoceros!" in the Milner Library Special Collections, Illinois State University, Normal). Nevertheless, Frost's later claim of 1872 in a candid letter to a business associate (in which he also indicated knowledge of the specimen in the London Zoo) should be accorded greater validity than an advertisement. See Richard J. Reynolds, "Circus Rhinos," *Bandwagon* 12 (November–December 1968): 4–13, for an excellent discussion that should be supplemented by Stuart Thayer, "One Sheet," *Bandwagon* 19 (May–June 1975): 31.

21. *A Spangled World; or, Life with the Circus* (New York, [1882]), 31; Olive Logan, *Before the Footlights and behind the Scenes* (Philadelphia, 1870), 316–19; Harry H. Marks, *Small Change; or, Lights and Shades of New York* (New York, 1882), 73–75; and Coup, *Sawdust and Spangles,* 20–26, 260.

22. R. G. Dun credit ledgers, New York, vol. 500, pp. 5, a/16, A-38, Baker Library, Harvard University Graduate School of Business Administration, Boston, Massachusetts.

23. Richard E. Conover, *The Great Forepaugh Show: America's Largest Circus from*

1864 (Xenia, Ohio: by the author, 1958); Richard P. Jones, *Sketches Pictorial and Descriptive of the Animals and Birds Contained in Forepaugh's Menagerie* (New York, 1867), 12, 33–34; Giles Pullman, *The American Kingdom Illustrated and Sketches Descriptive of the Beasts and Birds Contained in Forepaugh's Menagerie* (Buffalo, 1875), 11.

24. Logan, *Before the Footlights*, 314.

25. Conover, *Forepaugh*, 10; Stuart Thayer, *Mudshows and Railers: The American Circus in 1879* (Ann Arbor, Mich.: by the author, 1971), 39–40; Carl Hagenbeck, *Beasts and Men* (London: Longmans, Green, and Co., 1912), 26, 29.

26. Louis E. Cooke, "Reminiscences of a Showman," *Newark Evening Star*, June 17, 1915.

27. Henry D. Butler, *The Family Aquarium* (New York, 1858).

28. Coup, *Sawdust and Spangles*, 18–34, 247–62; R. G. Dun credit ledgers, New York, vol. 386, pp. 1500 A-21, 1314.

29. *New York Times*, October 6, 1876, 8; *New York Tribune*, October 9, 1876, 3; *New York Tribune*, October 28, 1876, 2; *New York Tribune*, December 1, 1876, 3.

30. *Guide to the New York Aquarium* (New York, 1877); *New York Times*, June 30, 1878, 5.

31. *Route of the W. C. Coup New United Monster Shows, New York Aquarium, Wonderful Broncho Horses, Royal Japanese Circus, Melville's Australian Circus, Colvin's Great Menagerie, Fryer's Startling Trained Animals and Wood's Museum, for the Traveling Season of 1879* (Lexington, Ky., 1879), 4.

32. *New York Times*, April 27, 1881, 8; see also the *New York Tribune*, February 24, 1881, 4.

33. Coup, *Sawdust and Spangles*, 20.

34. P. T. Barnum, *Struggles and Triumphs: Or, The Life of P. T. Barnum*, ed. George S. Bryan (New York: Knopf, 1927), 2:691; Hagenbeck, *Beasts and Men*, 11, where he errs in the year Barnum visited; the agents are listed in the show's published route books for 1877–80; see, for example, Charles H. White, *Statistics of P. T. Barnum's New and Greatest Show on Earth for the Season of 1877* (n.p., 1877), 3.

35. Carl Hagenbeck to Messrs. Ringling Brothers, October 7, 1902, collection of Fred D. Pfening III, Columbus, Ohio; Hagenbeck, *Beasts and Men*, 26.

36. Lorenz Hagenbeck, *Animals Are My Life* (London: Bodley Head, 1956), 16–19; Loisel, *Histoire*, 2:198; Adrienne Kaeppler, *Artificial Curiosities* (Honolulu: Bishop Museum, 1978), 14; W. Bulloch, *An Account of the Family of Laplanders . . . with their Summer and Winter Residences, Domestic Implements, Sledges* (London, [ca. 1822]).

37. *New York Times*, June 18, 1887, 2.

38. The intellectual framework is best discussed in Paul Greenhalgh, *Ephemeral Vistas* (Manchester: Manchester University Press, 1988), chap. 4, "Human Showcases," 82–111. Further details about the exhibitions of the Jardin d'Acclimatation, including the quote by a member of the Paris Anthropological Society, appear in William H. Schneider, *An Empire for the Masses: The French Popular Image of Africa, 1817–1900* (Westport, Conn.: Greenwood Press, 1982), chap. 6, "Africans in Paris," 125–51. Neither Greenhalgh nor Schneider makes the connection with Carl Hagenbeck, whose numerous ethnographic importations are recounted in his *Beasts and Men*, 16–20 and passim. For additional information on the Jardin d'Acclimatation, see also the chapter by Michael A. Osborne in this volume.

39. The press agent's presentation of these features appears in an advertising "courier," *History of the $200,000 Sacred White Elephant, Ethnological Congress of Savage and Barbarous Tribes and Book of Jumbo* (Buffalo, [1884]); the season's route book compiled by S. S. Smith, *My Diary or Route Book of P. T. Barnum's Greatest Show on Earth, and the Great London Circus for the Season of 1884* (n.p., [1884]), 13–14, lists the personnel and cultural

groups presented. On the two later ethnological congresses, see Harvey L. Watkins, *The Barnum & Bailey Official Route Book. . . Season of 1894* (Buffalo, [1894]), 15–17, 27–28, and George E. Hardy, *The Barnum & Bailey Official Route Book . . . Season of 1895* (Buffalo, [1895]), 15–16, 29. Circus executive Louis E. Cooke, in his serialized "Reminiscences of a Showman" appearing in the *Newark Evening Star,* credited J. B. Gaylord with arranging for Barnum's white elephant as well as the ethnological congresses in installments appearing on June 17 and October 14, 1915.

40. Hagenbeck's Chicago fair enterprise, including the Indian Ocean aquarium, is described in a chromolithographed advertising flyer, *Hagenbeck's Zoological Arena* [1893] in the author's collection. Hagenbeck's 1896 patent is noted in Edward P. Alexander, *Museum Masters* (Nashville: American Association for State and Local History, 1983), 324. Although Hagenbeck is not mentioned, the ethnic displays are fully discussed in Robert Rydell, *All the World's a Fair: Visions of Empire at American International Expositions, 1876–1916* (Chicago: University of Chicago Press, 1984), 38–71.

41. Kenneth Hudson, *Museums of Influence* (Cambridge: Cambridge University Press, 1987), vii–viii, 77 ff.

The Origin and Development of American Zoological Parks to 1899

The primary information on zoological parks was obtained from a survey conducted by the author (Vernon N. Kisling Jr., "The History of Zoological and Botanical Collections: A Survey," 1985–89), as well as subsequent requests for institutional histories which the author has made on behalf of the AZA History Task Force, American Zoo and Aquarium Association. In addition to the survey itself, many zoological parks sent typescript information, as well as published articles and other information (these have been cited if there was an author). Dates for the establishment of the zoological parks are given in the directories of the American Zoo and Aquarium Association and the American Association of Museums; however, these dates do not always agree and may be different from dates given in other sources, so the author's surveys were used for determining dates.

1. John Hendley Barnhart, "An Account of the Two Hundredth Anniversary of the Founding of the First Botanic Garden in the American Colonies by John Bartram," *Bartonia,* Proceedings of the Philadelphia Botanical Club, special issue, supplement to no. 12 (1931); John T. Faris, *Old Gardens in and about Philadelphia and Those Who Made Them* (Indianapolis: Bobbs-Merrill, 1932); John W. Harshberger, *The Botanists of Philadelphia and Their Work* (Philadelphia, 1899).

2. Paul M. Rea, "One Hundred and Fifty Years of Museum History," *Science* 57 (June 15, 1923): 677–81; Charles Coleman Sellers, *Mr. Peale's Museum: Charles Willson Peale and the First Popular Museum of Natural Science and Art* (New York: W. W. Norton, 1980).

3. George C. D. Odell, *Annals of the New York Stage* (New York: Columbia University Press, 1927); R. W. G. Vail, *Random Notes on the History of the Early American Circus* (Barre, Mass.: Barre Gazette, 1956).

4. Stuart Thayer, letters to author, 1988–89.

5. Arthur F. W. Hughes, *The American Biologist through Four Centuries* (Springfield, Ill.: Charles C. Thomas, 1982); William Martin Smallwood, *Natural History and the American Mind* (1941; New York: AMS Press, 1967); Raymond Phineas Stearns, *Science in the British Colonies of America* (Chicago: University of Illinois Press, 1970); Vernon N. Kisling Jr., "Zoological Curiosities from the Americas in Sixteenth- to Nineteenth-Century Menageries," in *The Exploration and Opening Up of America As Mirrored by Natural History* (Vienna: Naturhistorisches Museum Wien, 1992 conference proceedings in press).

6. U.S. Department of Commerce, *Historical Statistics of the United States: Colonial Times to 1970* (Washington: Government Printing Office, 1975).

7. Hughes, *American Biologist*; Smallwood, *Natural History*; Stearns, *Science in the British Colonies.*

8. Odell, *Annals of the New York Stage*; Vail, *Random Notes.*

9. Ralph S. Bates, *Scientific Societies in the United States,* 3d ed. (Cambridge: MIT Press, 1965), 5–6; Murphy D. Smith, *Oak from an Acorn: A History of the American Philosophical Society Library, 1770–1803* (Wilmington, Del.: Scholarly Resources, 1976), 1–3.

10. Barnhart, "Account of the Two Hundredth Anniversary"; Faris, *Old Gardens*; Harshberger, *Botanists of Philadelphia.*

11. Rea, "One Hundred and Fifty Years"; Sellers, *Mr. Peale's Museum.*

12. George H. Daniels, *American Science in the Age of Jackson* (New York: Columbia University Press, 1968); George H. Daniels, *Science in American Society: A Social History* (New York: Knopf, 1971); John C. Greene, *American Science in the Age of Jefferson* (Ames: Iowa State University Press, 1984); Brooke Hindle, *The Pursuit of Science in Revolutionary America, 1735–1789* (Chapel Hill: University of North Carolina Press, 1956); Hughes, *American Biologist*; Smallwood, *Natural History*; Stearns, *Science in the British Colonies.*

13. Roderick Nash, *Wilderness and the American Mind,* 3d ed. (New Haven: Yale University Press, 1982); Keith Thomas, *Man and the Natural World: A History of the Modern Sensibility* (New York: Pantheon Books, 1983).

14. Odell, *Annals of the New York Stage*; Vail, *Random Notes.*

15. Thayer, letters to author, 1988–89; Vail, *Random Notes.*

16. E. H. Bostock, *Menageries, Circuses, and Theaters* (1927; New York: Benjamin Blom, 1972); George L. Chindahl, *A History of the Circus in America* (Caldwell, Idaho: Caxton Printers, 1959); Wilton Eckley, *The American Circus* (Boston: Twayne, 1984); Richard W. Flint's chapter in this volume; Joanne C. Joys, *The Wild Animal Trainer in America* (Boulder, Colo.: Pruett, 1983); Earl Chapin May, *The Circus from Rome to Ringling* (1932; New York: Dover, 1963); Vail, *Random Notes.*

17. Robert V. Bruce, *The Launching of Modern American Science, 1846–1876* (New York: Knopf, 1987); A. Hunter Dupree, *Science in the Federal Government: A History of Policies and Activities* (1957; Baltimore: Johns Hopkins University Press, 1986); Walter B. Hendrickson, "Nineteenth-Century State Geological Surveys: Early Government Support of Science," *Isis* 52, no. 3 (1961): 357–71; William Stanton, *The Great United States Exploring Expedition of 1838–1842* (Berkeley: University of California Press, 1975).

18. James Fisher, *Zoos of the World: The Story of Animals in Captivity* (Garden City, N.Y.: Natural History Press, 1967); Rosl Kirchshofer, ed., *The World of Zoos: A Survey and Gazetteer,* trans. Hilda Morris (New York: Viking, 1968); Gustave Loisel, *Histoire des ménageries de l'antiquité à nos jours,* 3 vols. (Paris: Octave Doin et Fils and Henri Laurens, 1912); Doris Rybot, *It Began before Noah* (London: Michael Joseph, 1972).

19. Wilfrid Blunt, *The Ark in the Park: The Zoo in the Nineteenth Century* (London: Hamish Hamilton and the Tryon Gallery, 1976); P. Chalmers Mitchell, *Centenary History of the Zoological Society of London* (London: Zoological Society of London, 1929); Lord Zuckerman, ed., *The Zoological Society of London, 1826–1976 and Beyond,* Symposia of the Zoological Society of London, No. 40 (London: Zoological Society of London and Academic Press, 1976).

20. Mitchell, *Centenary History.*

21. *Oxford English Dictionary,* 2d ed.

22. Smithsonian Institution, *Annual Report of the Board of Regents of the Smithsonian Institution, Showing the Operations, Expenditures, and Condition of the Institution for the Year Ending June 30, 1897: Report of the U.S. National Museum, Part II* (Washington: Government Printing Office, 1901); Joel R. Poinsett, *Discourse on the Objects and Importance*

of the *National Institution for the Promotion of Science, Established at Washington, 1840* (Washington, 1841).

23. William Jones Rhees, ed., *The Smithsonian Institution: Documents Relative to Its Origin and History, 1835–1899* (Washington: Government Printing Office, 1901).

24. J. Thomas Scharf and Thompson Westcott, *History of Philadelphia, 1609–1884* (Philadelphia, 1884).

25. Zoological Society of Philadelphia, *An Act to Incorporate the Zoological Society of Philadelphia* (Philadelphia, 1859), 4.

26. Alyssa N. Scheuermann, "'Firsts' at the Zoological Society of Philadelphia," typescript, Zoological Society of Philadelphia, 1984, 3.

27. Williams Biddle Cadwalader, *Bears, Owls, Tigers, and Others! Philadelphia's Zoo, 1874–1949* (New York: Newcomen Society in North America, 1949), 11; Scharf and Westcott, *History of Philadelphia,* 1867–68; T. B. White, *Fairmount, Philadelphia's Park: A History* (Philadelphia: Art Alliance Press, 1975), 46.

28. Cadwalader, *Bears, Owls, Tigers,* 13; White, *Fairmount,* 54; Zoological Society of Philadelphia, *An Animal Garden in Fairmount Park* (Philadelphia: Zoological Society of Philadelphia, 1988), 7.

29. White, *Fairmount,* 54.

30. Ibid., 56; Scharf and Westcott, *History of Philadelphia,* 1863.

31. Cadwalader, *Bears, Owls, Tigers,* 15; Zoological Society of Philadelphia, *Animal Garden,* 14–16.

32. Cadwalader, *Bears, Owls, Tigers;* Vernon N. Kisling, "The History of Zoological and Botanical Collections: A Survey," 1985–89 (survey data and correspondence); Scharf and Westcott, *History of Philadelphia;* Scheuermann, "'Firsts' at the Zoological Society of Philadelphia"; Alyssa N. Scheuermann, "Zoo Firsts: Important Events in the History of the Zoological Society of Philadelphia," typescript, Zoological Society of Philadelphia, 1986; White, *Fairmount;* Zoological Society of Philadelphia, *Centennial Celebration, 1859–1959* (Philadelphia: Zoological Society of Philadelphia, 1959); Zoological Society of Philadelphia, *Animal Garden.*

33. William Bridges, *Gathering of Animals: An Unconventional History of the New York Zoological Society* (New York: Harper and Row, 1974); Kisling, "History of Zoological and Botanical Collections"; John W. Smith, "Central Park Animals As Their Keeper Knows Them," *Outing* 42 (May 1903): 248–54.

34. Kisling, "History of Zoological and Botanical Collections"; Mark A. Rosenthal, "It Began with a Gift," *Ark* 4, no. 3 (1976–77): 2–5, 12.

35. Oliver M. Gale, "The Cincinnati Zoo: One Hundred Years of Trial and Triumph," *Cincinnati Historical Society Bulletin* 33, no. 2 (1975): 89.

36. Ibid., 94.

37. Gale, "Cincinnati Zoo"; Kisling, "History of Zoological and Botanical Collections"; David Ehrlinger, *The Cincinnati Zoo and Botanical Garden: From Past to Present* (Cincinnati: Cincinnati Zoo and Botanical Garden, 1993).

38. Mildred F. Heap, *The Buffalo Zoo Story* (Buffalo: Buffalo Zoological Gardens, 1982).

39. Ibid., 12.

40. Heap, *Buffalo Zoo Story;* Kisling, "History of Zoological and Botanical Collections."

41. Laws of Maryland, chap. 344, Charter for the Baltimore Zoo, 1876.

42. Kisling, "History of Zoological and Botanical Collections."

43. Margaret Corell, "1882 . . . Beginning of the Zoo: The Early Years," *Zoo News* (winter 1981–82): 1–12; Kisling, "History of Zoological and Botanical Collections"; Charles

R. Voracek, "Zoo History, the Middle and Later Years, 1940–1960," *Zoo News* (spring 1982): 1–7.

44. R. Jeffrey Stott, "The American Idea of a Zoological Park: An Intellectual History" (Ph.D. diss., University of California, Santa Barbara, 1981).

45. "Zoo versus Menagerie," *Living Age* 317 (May 19, 1923): 375; Theodore Link, "Zoological Gardens: A Critical Essay," *American Naturalist* 17, no. 12 (1883): 1225–29; Henry S. Salt, *Animals' Rights: Considered in Relation to Social Progress* (1892; Clarks Summit, Pa.: Society for Animals' Rights, 1980).

46. P. T. Barnum, *Struggles and Triumphs: Or, The Life of P. T. Barnum,* ed. George S. Bryan (New York: Knopf, 1927); John Rickards Betts, "P. T. Barnum and the Popularization of Natural History," *Journal of the History of Ideas* 20, no. 3 (1959): 353–68.

47. Rhees, *Smithsonian Institution,* 685; Washington Zoological Society, "Charter, &c. of Washington Zoological Society," 1872, Smithsonian Institution Archives, Record Unit 7002, Spencer F. Baird Collection, Washington, D.C.

48. Smithsonian Institution, *Annual Report of the Board of Regents of the Smithsonian Institution, Showing the Operations, Expenditures, and Condition of the Institution for the Year Ending June 30, 1888: Report of the U.S. National Museum* (Washington: Government Printing Office, 1890), 20.

49. 25 Stat. 808.

50. Samuel Pierpont Langley, *Annual Report of the Smithsonian Institution, Year ending June 30, 1889* (Washington, 1889); Smithsonian Institution Archives, October 31, 1891, Records Unit 74.

51. Alfred Meyer, ed., *A Zoo for All Seasons* (New York: W. W. Norton, 1979), 34.

52. Julie Hamman, *Guide to Animal-Related Records at the National Zoological Park, 1887–1985,* Occasional Papers No. 1 (Washington: Smithsonian Institution Archives, 1988); Kisling, "History of Zoological and Botanical Collections"; Rhees, *Smithsonian Institution;* Smithsonian Institution, *Annual Report of the Board of Regents of the Smithsonian Institution, Showing the Operations, Expenditures, and Condition of the Institution to July, 1888* (Washington: Government Printing Office, 1890); Smithsonian Institution, *Annual Report of the Board of Regents of the Smithsonian Institution, Showing the Operations, Expenditures, and Condition of the Institution for the Year Ending June 30, 1888;* Smithsonian Institution, *Annual Report of the Board of Regents of the Smithsonian Institution, Showing the Operations, Expenditures, and Condition of the Institution to July, 1889* (Washington: Government Printing Office, 1890); Smithsonian Institution, *Annual Report of the Board of Regents of the Smithsonian Institution, Showing the Operations, Expenditures, and Condition of the Institution to July, 1890* (Washington: Government Printing Office, 1891); Webster Prentiss True, *The Smithsonian Institution,* Smithsonian Scientific Series, vol. 1 (1929; New York: Smithsonian Institution Series, 1943).

53. Richard J. Reynolds, "History of the Atlanta Zoo," in *Atlanta's Zoo* (Atlanta: City of Atlanta, 1969).

54. Kisling, "History of Zoological and Botanical Collections."

55. Ibid.

56. New York Zoological Society, "The New York Aquarium," *Zoological Society Bulletin* 8 (January 1903): 70–72; Kisling, "History of Zoological and Botanical Collections."

57. Ray Bamrick, "The Pittsburgh Zoo: A Look Back," *Animal Talk* (spring 1988): 17–19; Kisling, "History of Zoological and Botanical Collections."

58. Kisling, "History of Zoological and Botanical Collections."

59. Bridges, *Gathering of Animals,* 2.

60. Douglas J. Preston, *Dinosaurs in the Attic: An Excursion into the American Museum of Natural History* (New York: St. Martin's Press, 1986).

61. New York Zoological Society, *First Annual Report of the New York Zoological Society* (New York: New York Zoological Society, 1897), 52.

62. Ibid., 13.

63. Vernon N. Kisling Jr., "Libraries and Archives in the Historical and Professional Development of American Zoological Parks," *Libraries and Culture* 28, no. 3 (summer 1993): 247–65.

64. Michael Robinson, "Beyond the Zoo: The Biopark," *Defenders* 62, no. 6 (1987): 10–17.

The National Zoological Park

I am grateful for the unpublished history of the National Zoological Park, particularly those sections on construction phases compiled by the late Sybil E. (Billie) Hamlet, Office of Public Affairs, National Zoological Park, Washington, D.C. I would also like to thank the Smithsonian Institution for the Fellowship in American and Cultural History, which supported me while I researched and wrote the original article on which this chapter is based.

1. William Jones Rhees, ed., *The Smithsonian Institution: Documents Relative to Its Origin and History, 1835–1899* (Washington: Government Printing Office, 1901).

2. John Durant and Alice Durant, *Pictorial History of the American Circus* (New York: A. J. Barnes, 1957), 50–185 passim.

3. Sources for the history of zoos in Europe and America include James Fisher, *Zoos of the World*, ed. M. H. Chandler and Vernon Reynolds (London: Aldus Books, 1966); Emily Hahn, *Zoos* (London: Camelot Press, 1968); Rosl Kirchshofer, ed., *The World of Zoos: A Survey and Gazetteer*, trans. Hilda Morris (London: B. T. Batsford, 1968); Gustave Loisel, *Histoire des ménageries de l'antiquité à nos jours*, 3 vols. (Paris: Octave Doin et Fils and Henri Laurens, 1912); Charles V. A. Peel, *The Zoological Gardens of Europe: Their History and Chief Features* (London: F. E. Robinson and Co., 1903).

4. *Album of the Zoological Garden of Cincinnati* (Cincinnati, 1878).

5. Arthur Erwin Brown, *Guide to the Garden of the Zoological Society of Philadelphia* (Philadelphia, 1878).

6. Wilcomb E. Washburn, "Joseph Henry's Conception of the Purpose of the Smithsonian Institution," in *A Cabinet of Curiosities: Five Episodes in the Evolution of American Museums* (Charlottesville: University Press of Virginia, 1967), 106–66.

7. Samuel Pierpont Langley, "Memoir of George Brown Goode, 1851–1896," in *A Memorial of George Brown Goode, Annual Report of the Board of Regents of the Smithsonian Institution, June 30, 1897* (Washington: Government Printing Office, 1901), 2:46.

8. George Brown Goode, "The Principles of Museum Administration," in *A Memorial of George Brown Goode, Annual Report of the Board of Regents of the Smithsonian Institution, June 30, 1897* (Washington, 1897).

9. David Starr Jordan, "George Brown Goode," in *Dictionary of American Biography* (New York: Charles Scribner's Sons, 1931), 7:381–82.

10. Fairfield Osborn, "William Temple Hornaday," in *Dictionary of American Biography*, supplement 2 (New York: Charles Scribner's Sons, 1958), 316–18; John Ripley Forbes, *In the Steps of the Great American Zoologist: William Temple Hornaday* (New York: M. Evans and Co., 1966).

11. William T. Hornaday, "The Extermination of the American Bison with a Sketch of Its Discovery and Life History," in *Report of the United States National Museum under the Direction of the Smithsonian Institution* (Washington, 1889), 391.

12. June 27, 1887, Smithsonian Institution Archives, Record Unit 74.

13. Samuel Pierpont Langley, *Annual Report of the Smithsonian Institution, Year ending*

June 30, 1889 (Washington, 1890), 28. Hereafter, only the year of a regular annual report will be given.

14. Speech by William C. P. Breckinridge, April 9, 1890, quoted in Rhees, *Smithsonian Institution: Documents, II,* 1376–79.

15. See, for example, speech of M. A. Foran, September 12, 1888, quoted in Rhees, *Smithsonian Institution: Documents, II,* 1161–64.

16. *Annual Report of the Smithsonian Institution . . . 1893,* 27.

17. Speech by Joseph G. Cannon, February 5, 1891, quoted in Rhees, *Smithsonian Institution: Documents, II,* 1436.

18. Speech by J. H. Reagan, February 24, 1891, quoted in Rhees, *Smithsonian Institution: Documents, II,* 1463–64.

19. Langley, *Annual Report of the Smithsonian Institution . . . 1889,* 28.

20. October 31, 1891, Smithsonian Institution Archives, Record Unit 74.

21. December 10, 1890, Smithsonian Institution Archives, Record Unit 74.

22. Samuel Pierpont Langley, October 31, 1890, Smithsonian Institution Archives, Record Unit 74.

23. William T. Hornaday, April 3, 1890, Smithsonian Institution Archives, Record Unit 74, 181–83.

24. William T. Hornaday, January 9, 1890, Smithsonian Institution Archives, Record Unit 74, 164, 96–98.

25. David Hancocks, *Animals and Architecture* (London: H. Evelyn, 1971), esp. 110–11.

26. *Annual Report of the Smithsonian Institution . . .* 1891, 21–23; *Annual Report of the Smithsonian Institution . . . 1892,* 28–45.

27. Samuel Pierpont Langley, March 4, 1891, Smithsonian Institution Archives, Record Unit 74.

28. Ibid.

29. *Annual Report of the Smithsonian Institution . . . 1891,* 50.

30. Frank Baker, November 13, 1893, Smithsonian Institution Archives, Record Unit 74, 677.

31. Samuel Pierpont Langley, March 4, 1891, Smithsonian Institution Archives.

32. Samuel Pierpont Langley, May 7, 1890, Smithsonian Institution Archives, Record Unit 34; Samuel Pierpont Langley, January 3, 1891, Smithsonian Institution Archives, Record Unit 74.

33. William T. Hornaday, May 16, 1890, Smithsonian Institution Archives, Record Unit 7003.

34. Cyrus Adler, "Samuel Pierpont Langley," in *Annual Report of the Smithsonian Institution . . . 1906,* 531–32; J. G. Vaeth, *Langley: Man of Science and Flight* (New York: Ronald Press Co., 1966), 54–62.

35. William T. Hornaday, March 3–8, 1890, Smithsonian Institution Archives, Record Unit 74.

36. William T. Hornaday, April 17, 1890, Smithsonian Institution Archives, Record Unit 74.

37. Samuel Pierpont Langley, May 29, 1890, Smithsonian Institution Archives, Record Unit 74.

38. Samuel Pierpont Langley, April 30, 1892, Smithsonian Institution Archives, Record Unit 74.

39. Frank Baker, April 28, 1892, Smithsonian Institution Archives, Record Unit 74; Samuel Pierpont Langley, June 17, 1892, Smithsonian Institution Archives, Record Unit 74.

40. Frank Baker, August 26, 1892, Smithsonian Institution Archives, Record Unit 74, 32.

41. Samuel Pierpont Langley, November 23, 1893, Smithsonian Institution Archives, Record Unit 74.

42. Frank Baker, October 29, 1890, Smithsonian Institution Archives, Record Unit 74, 23-24; Frank Baker, August 25, 1890, Smithsonian Institution Archives, Record Unit 74, 370-76.

43. *Annual Report of the Smithsonian Institution . . . 1892*, 70.

44. Frank Baker, "The National Zoological Park," in *The Smithsonian Institution, 1846-1896: The History of the First Half Century,* ed. George Brown Goode (Washington: Government Printing Office, 1897); Frank Baker, January 7, 1891, Smithsonian Institution Archives, Record Unit 74, 280-82.

45. Sybil E. Hamlet, unpublished history of the National Zoological Park.

46. Samuel Pierpont Langley, March 23, 1895, Smithsonian Institution Archives, Record Unit 74.

47. Samuel Pierpont Langley, December 20, 1894, Smithsonian Institution Archives, Record Unit 74.

48. Samuel Pierpont Langley, January 22, 1891, Smithsonian Institution Archives, Record Unit 74; Samuel Pierpont Langley, December 21, 1893, Smithsonian Institution Archives, Record Unit 74.

49. Samuel Pierpont Langley, March 23, 1895, Smithsonian Institution Archives, Record Unit 74.

Epilogue

1. In Hagenbeck's words (translated from German): "I desired, above all things, to give the animals a maximum of liberty. I wished to exhibit them not as captives . . . but free to wander from place to place within as large limits as possible, and with no bars to obstruct the view and serve as a reminder of captivity . . . [thus] giving the animals as much freedom and placing them in as natural an environment as possible. . . . For the carnivores, glens had to be established, not confined with railings, but separated from the public with deep trenches, large enough to prevent the animals from getting out, but not in anyway interfering with the view." Carl Hagenbeck, *Beasts and Men,* abridged and translated by Hugh S. R. Elliot and A. G. Thacker, with an introduction by P. Chalmers Mitchell (London: Longmans, Green and Co., 1909), 40-42.

The Value of Old Photographs of Zoological Collections

1. H. Hediger, *Man and Animal in the Zoo,* trans. Gwynne Vevers and Winwood Reade (London: Routledge and Kegan Paul, 1969).

2. E. Mohr, *Das Urwildpferd (Equus przewalskii)* (Poljakoff, 1881; Wittenberg-Lutherstadt: A. Ziemsen, 1959).

3. R. A. Patlen, "'Jessie' Joins Her Ancestors," *Parks and Recreation* 23, no. 5 (1940): 200-202.

4. C. V. A. Peel, *The Zoological Gardens of Europe: Their History and Chief Features* (London: F. E. Robinson and Co., 1903).

The Architecture of the National Zoological Park

1. Samuel Pierpont Langley to Senator Samuel Dibble, January 18, 1889, Smithsonian Institution Archives, Record Unit 74 (National Zoological Park, 1887-1965, Records), Box 6.

2. William H. Pierson Jr., *American Buildings and Their Architects* (New York: Anchor Press, 1976), 14.

3. William Truettner, *National Parks and the American Landscape* (Washington: Smithsonian Institution Press, 1972).

4. Albert Fein, *Frederick Law Olmsted and the American Environmental Tradition* (New York: G. Braziller, 1972), 9.

5. Samuel Pierpont Langley, *Annual Report of the Board of Regents of the Smithsonian Institution Showing the Operations, Expenditures, and Conditions of the Institution to July, 1894* (Washington, 1896). Hereafter, this source will be cited as *Annual Report* with its year.

6. William Temple Hornaday, May 24, 1888, interview in the *Washington Herald*, "A Capital Zoo," Smithsonian Institution Archives, Record Unit 7081 (William J. Rhees Collection), Box 29; William Temple Hornaday, "Extermination of the American Bison," reprinted in the *Annual Report of the United States National Museum for 1887* (Washington, 1888), 367–548.

7. See Helen Lefkowitz Horowitz's chapter in this volume.

8. Samuel Pierpont Langley to Frederick Law Olmsted, to William Ralph Emerson, May 5, 1890, Smithsonian Institution Archives, Record Unit 34 (Office of the Secretary, 1887–1907, Outgoing Correspondence), vol. 5.1, 1, 2–3.

9. Samuel Pierpont Langley to George Brown Goode, May 7, 1890, Smithsonian Institution Archives, Record Unit 34, vol. 5.1, 9–12.

10. Frederick Law Olmsted Sr. to Samuel Pierpont Langley, May 7, 1890, Smithsonian Institution Archives, Record Unit 7081, Box 45.

11. Bruce Kelly, *Art of the Olmsted Landscape* (New York: Landmarks Preservation Commission, 1981), 19.

12. John C. Olmsted to Frederick Law Olmsted Sr., September 24, 1891, Library of Congress, Manuscript Division, Records of the Olmsted Associates, Job #2822, Box 134.

13. Samuel Pierpont Langley to Frank Baker, February 20, 1902, Smithsonian Institution Archives, Record Unit 74, Box 111.

14. Ibid.

15. Frank Baker, "The National Zoological Park," in George Brown Goode, *The Smithsonian Institution: 1846–1896, The History of Its First Half Century* (Washington, D.C., 1897), 455.

16. Frederick Law Olmsted Jr., memorandum, March 27, 1902, Library of Congress, Manuscript Division, Records of the Olmsted Associates, Job #2822, Box 134.

17. Memorandum Dictated by the Secretary [Samuel P. Langley] at the National Zoological Park, January 23, 1902, Smithsonian Institution Archives, Record Unit 34, vol. 5.7, 389.

18. Vincent Scully, *The Shingle Style* (New Haven: Yale University Press, 1955).

19. Obituary of William Ralph Emerson, *Journal of the American Institute of Architects* 6 (1918): 89.

20. Cynthia Zaitzevsky, *Architecture of William Ralph Emerson, 1833–1917* (Cambridge: Harvard University Press, 1969), 25–26.

21. William Ralph Emerson, "Free Hand Drawing," *Technology Architectural Review*, MIT Department of Architecture, 2, no. 6 (1889).

22. Samuel Pierpont Langley to William Ralph Emerson, October 12, 1899, Smithsonian Institution Archives, Record Unit 74, Box 11.

23. George L. Hersey, *High Victorian Gothic: A Study in Associationism* (Baltimore: Johns Hopkins University Press, 1972), 7.

24. Memorandum of meeting with Arthur Brown (in attendance: Langley, Baker, and Olmsted), June 18, 1890, Smithsonian Institution Archives, Record Unit 74, Box 289.

25. Roland Baetens, *The Chant of Paradise: The Antwerp Zoo: 150 Years of History* (Tielt, Belgium: Antwerp Zoo, 1993), 147.

26. Frank Baker to William Ralph Emerson, November 24, 1890, Smithsonian Institution Archives, Record Unit 74, Box 7.

---

27. Samuel Pierpont Langley to William Ralph Emerson, December 9, 1890, Smithsonian Institution Archives, Record Unit 34, vol. 5.1, 186.

28. Samuel Pierpont Langley to William Ralph Emerson, April 13, 1894, Smithsonian Institution Archives, Record Unit 34, Box 25.

29. William Ralph Emerson to Frederick Law Olmsted Jr., August 6, 1894, Library of Congress, Manuscript Division, Records of the Olmsted Associates, Job #2822, Box 134.

30. Frank Baker to Glenn Brown, May 7, 1899, Smithsonian Institution Archives, Record Unit 74, Box 10.

31. William Ralph Emerson to Samuel Pierpont Langley, May 10, 1899, Smithsonian Institution Archives, Record Unit 31 (Office of the Secretary, Incoming Correspondence), Box 24.

32. Frank Baker to Samuel Pierpont Langley, September 9, 1899, Smithsonian Institution Archives, Record Unit 74, Box 128.

33. Richard Rathbun to Hornblower and Marshall, August 18, 1902, Smithsonian Institution Archives, Record Unit 74, Box 128.

34. Cynthia R. Field, Richard E. Stamm, and Heather P. Ewing, *The Castle: An Illustrated History of the Smithsonian Building* (Washington, D.C.: Smithsonian Institution Press, 1993), 46, 126-30.

35. Samuel Pierpont Langley to Frank Baker, November 25, 1901, Smithsonian Institution Archives, Record Unit 74, Box 131.

36. Olmsted Associates to Frank Baker, April 20, 1894, Library of Congress, Manuscript Division, Records of the Olmsted Associates, Job #2822, Box 134.

37. Charles Moore to Frederick Law Olmsted Jr., May 3, 1902, Library of Congress, Manuscript Division, Records of the Olmsted Associates, Job #2822, Box 134.

38. Frank Baker to Hornblower and Marshall, March 4, 1903, Smithsonian Institution Archives, Record Unit 74, Box 13, 468-69.

39. Hornblower and Marshall to Frank Baker, March 25, 1903, Smithsonian Institution Archives, Record Unit 74, Box 126.

40. Frederick Law Olmsted Jr., April 9, 1903, Smithsonian Institution Archives, Record Unit 74, Box 126.

41. Ibid.

42. Commissioner of the Bureau of Fisheries to Frank Baker, April 24, 1905, Smithsonian Institution Archives, Record Unit 74, Box 126.

43. Frank Baker to Samuel Pierpont Langley, January 3, 1905, Smithsonian Institution Archives, Record Unit 74, Box 126.

44. Hornblower and Marshall to Perth Amboy Terra Cotta Company, December 12, 1904, Smithsonian Institution Archives, Record Unit 74, Box 126.

CONTRIBUTORS

WILLIAM A. DEISS Associate Archivist, Smithsonian Institution

JOHN C. EDWARDS Fellow of the Zoological Society of London and Chairman of the Bartlett Society, London

HEATHER EWING Office of Architectural History and Historic Preservation, Smithsonian Institution

RICHARD W. FLINT Baltimore, Maryland

LINDEN GILLBANK Department of History and Philosophy of Science, University of Melbourne

R. J. HOAGE, ANNE ROSKELL, AND JANE MANSOUR National Zoological Park, Smithsonian Institution

HELEN LEFKOWITZ HOROWITZ Professor of History and American Studies, Smith College

VERNON N. KISLING JR. Collection Management Coordinator, Marston Science Library, University of Florida

SALLY GREGORY KOHLSTEDT Professor of History of Science, University of Minnesota

D. K. MITTRA Deputy Librarian, National Library, Calcutta, India

MICHAEL A. OSBORNE Associate Professor of History and Environmental Studies, University of California at Santa Barbara

HERMAN REICHENBACH Science Section, Central Research Department, Gruner and Jahr (Newspapers and Magazines) AG & Co., Hamburg, Germany

HARRIET RITVO Professor of History, Massachusetts Institute of Technology

MICHAEL H. ROBINSON Director, National Zoological Park, Smithsonian Institution

HARRO STREHLOW Biologist, Zoo Historian, Author, Berlin, Germany

THOMAS VELTRE Manager, Media Services, Wildlife Conservation Society, Bronx, New York

INDEX

Zoological Institute, 98–99, 112
Zoological Society of Cincinnati, 118
Zoological Society of Hamburg, 52, 61
Zoological Society of London, 27, 43–47, 92, 148–49, 150; controversy over acquisitions, 44–46
Zoological Society of Victoria, 80–82
Zoologischer Garten bei Berlin. *See* Zoological Garden near Berlin
zoos: in Alexandria, 10; architecture of, 37, 66–69, 136–37, 141, 155–64; birth of, in America, 115–19, 127; birth of, in Europe, 113–14, 126–27; changing role of, 6–7, 27–28, 137; classi-cal, 21; development of, 21–22, 113–15; and electronic culture, 28–29; and endangered species, 120, 128, 137, 152–53; establishment of, 15–16; history of, 6–7, 8–18; in modern era, 122–25; as multifunctional institution, 38; nepotism in, 33; old photographs of, 141–50; in Paris, 33–42; in print era, 27–28; as scientific resource, 36–37, 38–39, 41, 46, 65, 92, 116, 122, 135, 151; symbolic role of, 28–29; in Victorian England, 43–50. *See also* menageries
Zuckerman, Lord, 8

Library of Congress Cataloging-in-Publication Data

New worlds, new animals : from menagerie to zoological park in the nineteenth century / edited by R. J. Hoage and William A. Deiss : with a foreword by Michael H. Robinson.
    p.   cm.
"Published in association with the National Zoological Park, Smithsonian Institution."
Most of the chapters are derived from papers delivered at a symposium held Oct. 1989 to celebrate the centennial of the Smithsoninan Institution's National Zoological Park.
    Includes bibliographical references (p.   ) and index.
    ISBN 0-8018-5110-6 (hc : alk. paper)—ISBN 0-8018-5373-7 (pbk : alk. paper)
    1. Zoo—History—19th century—Congresses. 2. Menageries—History—19th century—Congresses. 3. National Zoological Park (U.S.)—History—19th century—Congresses. I. Hoage, R. J. II. Deiss, William A. III. National Zoological Park (U.S.)
QL76.N48 1996
590'.74'409034—dc20   95-33006

Lightning Source UK Ltd.
Milton Keynes UK
UKOW05f1851250915

259285UK00004BA/188/P